Environmental Toxicology

Environmental Toxicology

Edited by
Kaiden Higgins

Larsen & Keller
www.larsen-keller.com

Environmental Toxicology
Edited by Kaiden Higgins
ISBN: 978-1-63549-113-5 (Hardback)

Larsen & Keller

Published by Larsen and Keller Education,
5 Penn Plaza,
19th Floor,
New York, NY 10001, USA

Cataloging-in-Publication Data

Environmental toxicology / edited by Kaiden Higgins.
    p. cm.
Includes bibliographical references and index.
ISBN 978-1-63549-113-5
    1. Environmental toxicology. 2. Pollutants--Toxicity testing. 3. Environmental health. I. Higgins, Kaiden.
RA1226 .E58 2017
615.902--dc23

The publisher's policy is to use permanent paper from mills that operate a sustainable forestry policy. Furthermore, the publisher ensures that the text paper and cover boards used have met acceptable environmental accreditation standards.

Printed and bound in the United States of America.

For more information regarding Larsen and Keller Education and its products, please visit the publisher's website www.larsen-keller.com

Table of Contents

Preface   **VII**

Chapter 1 **Introduction to Environmental Toxicology**   **1**

Chapter 2 **Environmental Pollution and its Sources**   **5**
    a. Bioaccumulation   5
    b. Toxic Heavy Metal   6
    c. Arsenic Contamination of Groundwater   13
    d. Mercury in Fish   19
    e. Lead Poisoning   31
    f. Cadmium Poisoning   54
    g. Oil Pollution Toxicity to Marine Fish   56
    h. Nuclear Fallout   60
    i. DDT   68
    j. Environmental Impact of Shipping   81
    k. Committed Dose   88
    l. Methylcyclopentadienyl Manganese Tricarbonyl   93
    m. Methylmercury   97
    n. Carbon Monoxide Poisoning   101
    o. Polycyclic Aromatic Hydrocarbon   112

Chapter 3 **Insecticides, Pesticides and Environmental Toxicology**   **124**
    a. Persistent Organic Pollutant   124
    b. Health Effects of Pesticides   132
    c. Pesticide Poisoning   135
    d. Toxicity Class   139
    e. Pesticide Toxicity to Bees   142

Chapter 4 **Significant Aspects of Environmental Toxicology**   **153**
    a. Environmental Hazard   153
    b. Ecological Death   156
    c. Persistent, Bioaccumulative and Toxic Substances   158
    d. Pathogen   163

Chapter 5 **Evaluation Methods of Toxicology**   **166**
    a. Air Quality Index   166
    b. Early Life Stage Test   179
    c. Modes of Toxic Action   181
    d. Bioanalysis   189
    e. Blood Lead Level   195
    f. Dietary Reference Intake   198

Chapter 6   **Study of Aquatic Toxicology**   **204**
            a. Aquatic Toxicology   204

Permissions

Index

Preface

Environmental toxicology refers to the branch of science which studies the damage caused to living organisms by different chemical, physical and biological agents. It includes ecotoxicology as its sub-field. It uses various methods to collect samples and assessments to ascertain the vulnerability of certain ecosystems and the best remedies to alleviate the situation. This book attempts to understand the multiple branches that fall under this discipline and how such concepts have practical applications. It presents this complex subject in the most comprehensible and easy to understand language. The topics covered in this text offer the readers new insights in the field of environmental toxicology. It aims to serve as a resource guide for students and experts alike and contribute to the growth of the discipline.

A foreword of all Chapters of the book is provided below:

Chapter 1 - This chapter introduces the reader to the discipline of environmental toxicology or entox as it is called for short, a multidisciplinary field of science that deals with the adverse effects of chemical, physical and biological agents on living organisms. It also explains about the scope and applications of entox. The chapter on environmental toxicology offers an insightful focus, keeping in mind the complex subject matter; **Chapter 2** - The advancement in science and technology came at the price of environmental degradation and pollution. This chapter of the book focuses on the various sources of environmental pollution, the far reaching impact of bioaccumulation, DDT, oil spills, nuclear fallout, carbon monoxide poisoning, lead poisoning and arsenic contamination of groundwater. This chapter elucidates the crucial theories and principles of environmental pollution and its sources; **Chapter 3** - The negative impact of using insecticides and pesticides on the environment has been well documented. Persistent organic pollutants like endrin, mirex, hexachlorobenzene and other such pesticides do not degrade easily and have really long half-lives that lead to their bioaccumulation. In this chapter, the reader is presented with information regarding the various pesticides and insecticides that damage the environment and are proven carcinogens and their impact on the flora and fauna; **Chapter 4** - Environmental toxicology deals largely with environmental hazards that are caused by the bioaccumulation of chemical substances. This accumulation of toxins results in ecological death of many organisms or in the impairment of the organism's ability to function. The role of pathogens cannot be undermined either; hence in this chapter we find an in-depth analysis of all these issues. The aspects elucidated in this chapter are of vital importance, and provide a better understanding of environmental toxicology; **Chapter 5** - To determine toxicity levels of the environment, there exist several measuring scales that enable a statistical understanding. This chapter of the book talks about scales and indexes like- dietary reference intake, air quality index, early life stage test etc. The chapter also studies the modes of toxic action and the quantification of toxic substances in living organisms. This chapter discusses the methods of environmental toxicology in a critical manner providing key analysis to the subject matter; **Chapter 6** - This chapter deals with the issue of toxicity rampant in aquatic organisms panning the freshwater, saltwater and sediment environments. The problem of bioaccumulation in aquatic organisms is a mammoth concern and this chapter provides extensive data on exposure systems, aquatic toxicity tests while also furnishing material on the toxicological effects of pollutants on aquatic organisms and how this indirectly affects human beings due to bioconcentration.

I would like to thank the entire editorial team who made sincere efforts for this book and my family who supported me in my efforts of working on this book. I take this opportunity to thank all those who have been a guiding force throughout my life.

Editor

Introduction to Environmental Toxicology

This chapter introduces the reader to the discipline of environmental toxicology or entox as it is called for short, a multidisciplinary field of science that deals with the adverse effects of chemical, physical and biological agents on living organisms. It also explains about the scope and applications of entox. The chapter on environmental toxicology offers an insightful focus, keeping in mind the complex subject matter.

Environmental toxicology, also known as entox, is a multidisciplinary field of science concerned with the study of the harmful effects of various chemical, biological and physical agents on living organisms. Ecotoxicology is a subdiscipline of environmental toxicology concerned with studying the harmful effects of toxicants at the population and ecosystem levels.

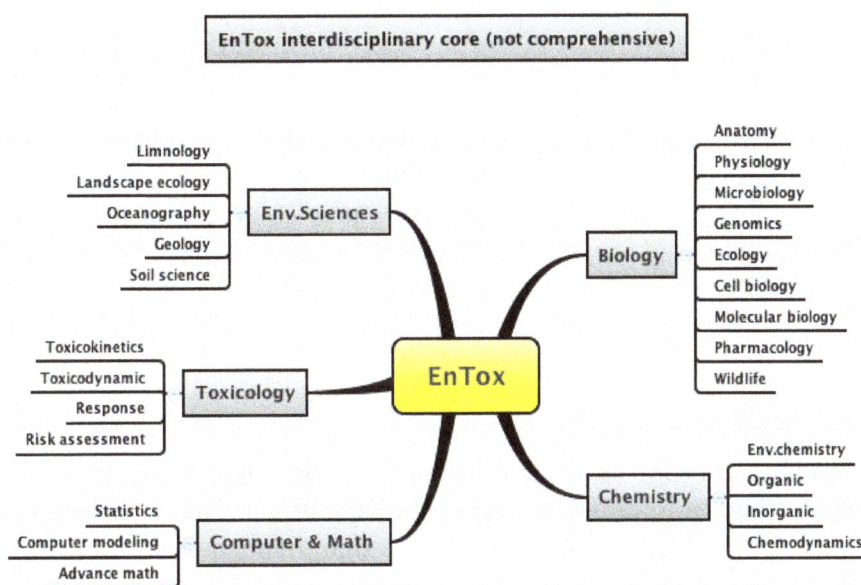

Overview of the interdisciplinarity of environmental toxicology

Rachel Carson is considered the mother of environmental toxicology, as she made it a distinct field within toxicology in 1962 with the publication of her book *Silent Spring*, which covered the effects of uncontrolled pesticide use. Carson's book was extensively based on a series of reports made by Lucille Farrier Stickel on the ecological effects of the pesticide DDT.

Organisms can be introduced to toxicants at various stages of their life cycle. The degree of toxicity can vary depending on where the organism is found within its food web. Bioaccumulation occurs when molecular compounds are stored in an organism's fatty tissues. Over time, this leads to the establishment of a trophic cascade and the biomagnification of specific toxicants. Biodegradation releases CO_2 and water as by-products into the environment. This process is typically limited in areas affected by environmental toxicants.

Harmful effects of chemical and biological agents can include toxicants from pollutants, insecticides, pesticides, and fertilizers, all of which can impact an organism and its community through shifts in species diversity and abundance. Resulting changes in population dynamics impact the ecosystem by altering its productivity and stability.

Legislation has been implemented since the early 1970s to ensure that harmful effects of environmental toxicants are minimized for all species. Unfortunately, according to McCarty (2013) we are facing the risk of entering in a "dark age" due to longstanding limitations in the implementation of the simple conceptual modes

Sources of Environmental Toxicity

There are many sources of environmental toxicity that can lead to the presence of toxicants in our food, water and air. These sources include organic and inorganic pollutants, pesticides and biological agents, all of which can have harmful effects on living organisms. There can be so called point sources of pollution, for instance the drains from a specific factory but also non-point sources (diffuse sources) like the rubber from car tires that contain numerous chemicals and heavy metals that are spread in the environment.

PCBs

Polychlorinated biphenyls (PCBs) are organic pollutants that are still present in our environment today, despite being banned in many countries, including the United States and Canada. Due to the persistent nature of PCBs in aquatic ecosystems, many aquatic species contain high levels of this chemical. For example, wild salmon (*Salmo salar*) in the Baltic Sea have been shown to have significantly higher PCB levels than farmed salmon as the wild fish live in a heavily contaminated environment.

Heavy Metals

Heavy metals found in food sources, such as fish can also have harmful effects. These metals can include mercury, lead and cadmium. It has been shown that fish (i.e. rainbow trout) are exposed to higher cadmium levels and grow at a slower rate than fish exposed to lower levels or none. Moreover, cadmium can potentially alter the productivity and mating behaviours of these fish. Heavy metals can not only affect behaviors, but also the genetic makeup in aquatic organisms. In Canada, a study examined genetic diversity in wild yellow perch along various heavy metal concentration gradients in lakes polluted by mining operations. Researchers wanted to determine as to what effect metal contamination had on evolutionary responses among populations of yellow perch. Along the gradient, genetic diversity over all loci was negatively correlated with liver cadmium contamination. Additionally, there was a negative correlation observed between copper contamination and genetic diversity. Some aquatic species have evolved heavy metal tolerances. In response to high heavy metal concentrations a Dipteran species, *Chironomus riparius*, of the midge family, *Chironomidae*, has evolved to become tolerant to Cadmium toxicity in aquatic environments. Altered life histories, increased Cd excretion, and sustained growth under Cd exposure is evidence that shows that Chironomus riparius exhibits genetically based heavy metal tolerance.

Pesticides

Pesticides are a major source of environmental toxicity. These chemically synthesized agents have been known to persist in the environment long after their administration. The poor biodegradability of pesticides can result in bioaccumulation of chemicals in various organisms along with biomagnification within a food web. Pesticides can be categorized according to the pests they target. Insecticides are used to eliminate agricultural pests that attack various fruits and crops. Herbicides target herbal pests such as weeds and other unwanted plants that reduce crop production.

DDT

Dichlorodiphenyltrichloroethane (DDT) is an organochlorine insecticide that has been banned due to its adverse effects on both humans and wildlife. DDT's insecticidal properties were first discovered in 1939. Following this discovery, DDT was widely used by farmers in order to kill agricultural pests such as the potato beetle, coddling moth and corn earworm. In 1962, the harmful effects of the widespread and uncontrolled use of DDT were detailed by Rachel Carson in her book The Silent Spring. Such large quantities of DDT and its metabolite Dichlorodiphenyldichloroethylene (DDE) that were released into the environment were toxic to both animals and humans.

DDT is not easily biodegradable and thus the chemical accumulates in soil and sediment runoff. Water systems become polluted and marine life such as fish and shellfish accumulate DDT in their tissues. Furthermore, this effect is amplified when animals who consume the fish also consume the chemical, demonstrating biomagnification within the food web. The process of biomagnification has detrimental effects on various bird species because DDT and DDE accumulate in their tissues inducing egg-shell thinning. Rapid declines in bird populations have been seen in Europe and North America as a result.

Humans who consume animals or plants that are contaminated with DDT experience adverse health effects. Various studies have shown that DDT has damaging effects on the liver, nervous system and reproductive system of humans.

By 1972, the United States Environmental Protection Agency (EPA) banned the use of DDT in the United States. Despite the regulation of this pesticide in North America, it is still used in certain areas of the world. Traces of this chemical have been found in noticeable amounts in a tributary of the Yangtze River in China, suggesting the pesticide is still in use in this region.

Sulfuryl Fluoride

Sulfuryl fluoride is an insecticide that is broken down into fluoride and sulfate when released into the environment. Fluoride has been known to negatively affect aquatic wildlife. Elevated levels of fluoride have been proven to impair the feeding efficiency and growth of the common carp (*Cyprinus carpio*). Exposure to fluoride alters ion balance, total protein and lipid levels within these fish, which changes their body composition and disrupts various biochemical processes.

References

- "Dioxins and PCBs report shows drop in dietary exposure over last decade | European Food Safety Authority". www.efsa.europa.eu. Retrieved 2016-02-04.

- "Lucille Farrier Stickel: Research Pioneer". National Wildlife Refuge System. United States Fish and Wildlife Service. March 7, 2014. Retrieved August 24, 2015.

- McCarty, L.S. (Dec 2013). "Are we in the dark ages of environmental toxicology?". Regul Toxicol Pharmacol. 67 (3): 321–324.

Environmental Pollution and its Sources

The advancement in science and technology came at the price of environmental degradation and pollution. This chapter of the book focuses on the various sources of environmental pollution, the far reaching impact of bioaccumulation, DDT, oil spills, nuclear fallout, carbon monoxide poisoning, lead poisoning and arsenic contamination of groundwater. This chapter elucidates the crucial theories and principles of environmental pollution and its sources.

Bioaccumulation

Bioaccumulation refers to the accumulation of substances, such as pesticides, or other chemicals in an organism. Bioaccumulation occurs when an organism absorbs a - possibly toxic - substance at a rate faster than that at which the substance is lost by catabolism and excretion. Thus, the longer the biological half-life of a toxic substance the greater the risk of chronic poisoning, even if environmental levels of the toxin are not very high. Bioaccumulation, for example in fish, can be predicted by models. Hypotheses for molecular size cutoff criteria for use as bioaccumulation potential indicators are not supported by data. Biotransformation can strongly modify bioaccumulation of chemicals in an organism.

Bioconcentration is a related but more specific term, referring to uptake and accumulation of a substance from water alone. By contrast, bioaccumulation refers to uptake from all sources combined (e.g. water, food, air, etc.)

Examples

An example of poisoning in the workplace can be seen from the phrase "as mad as a hatter" (18th and 19th century England). The process for stiffening the felt used in making hats more than a hundred years ago involved mercury, which forms organic species such as methylmercury, which is lipid-soluble, and tends to accumulate in the brain, resulting in mercury poisoning. Other lipid-soluble (fat-soluble) poisons include tetraethyllead compounds (the lead in leaded petrol), and DDT. These compounds are stored in the body's fat, and when the fatty tissues are used for energy, the compounds are released and cause acute poisoning.

Strontium-90, part of the fallout from atomic bombs, is chemically similar enough to calcium that it is utilized in osteogenesis, where its radiation can cause damage for a long time.

Naturally produced toxins can also bioaccumulate. The marine algal blooms known as "red tides" can result in local filter feeding organisms such as mussels and oysters becoming toxic; coral fish can be responsible for the poisoning known as ciguatera when they accumulate a toxin called ciguatoxin from reef algae.

Some animal species exhibit bioaccumulation as a mode of defense; by consuming toxic plants or animal prey, a species may accumulate the toxin, which then presents a deterrent to a potential predator. One example is the tobacco hornworm, which concentrates nicotine to a toxic level in its body as it consumes tobacco plants. Poisoning of small consumers can be passed along the food chain to affect the consumers later on. Other compounds that are not normally considered toxic can be accumulated to toxic levels in organisms. The classic example is of Vitamin A, which becomes concentrated in carnivore livers of e.g. polar bears: as a pure carnivore that feeds on other carnivores (seals), they accumulate extremely large amounts of Vitamin A in their livers. It was known by the native peoples of the Arctic that the livers of carnivores should not be eaten, but Arctic explorers have suffered Hypervitaminosis A from eating the bear livers (and there has been at least one example of similar poisoning of Antarctic explorers eating husky dog livers). One notable example of this is the expedition of Sir Douglas Mawson, where his exploration companion died from eating the liver of one of their dogs.

Coastal fish (such as the smooth toadfish) and seabirds (such as the Atlantic puffin) are often monitored for heavy metal bioaccumulation.

In some eutrophic aquatic systems, biodilution can occur. This trend is a decrease in a contaminant with an increase in trophic level and is due to higher concentrations of algae and bacteria to "dilute" the concentration of the pollutant.

Toxic Heavy Metal

A 25-foot (7.6 m) wall of coal fly ash contaminated with toxic heavy metals, resulting from the release of 5.4 million cubic yards of coal fly ash slurry into the Emory River, Tennessee, and nearby land and water features, in December 2008. Testing showed significantly elevated levels of arsenic, copper, barium, cadmium, chromium, lead, mercury, nickel, and thallium in samples of slurry and river water. Cleanup costs may exceed $1.2 billion.

A toxic heavy metal is any relatively dense metal or metalloid that is noted for its potential toxicity, especially in environmental contexts. The term has particular application to cadmium, mercury, lead and arsenic, all of which appear in the World Health Organisation's list of 10 chemicals of major public concern. Other examples include manganese, chromium, cobalt, nickel, copper, zinc, selenium, silver, antimony and thallium.

Heavy metals are found naturally in the earth. They become concentrated as a result of human caused activities and can enter plant, animal, and human tissues via inhalation, diet, and manual handling. Then, they can bind to and interfere with the functioning of vital cellular components. The toxic effects of arsenic, mercury, and lead were known to the ancients, but methodical studies of the toxicity of some heavy metals appear to date from only 1868. In humans, heavy metal poisoning is generally treated by the administration of chelating agents. Some elements otherwise regarded as toxic heavy metals are essential, in small quantities, for human health.

Contamination Sources

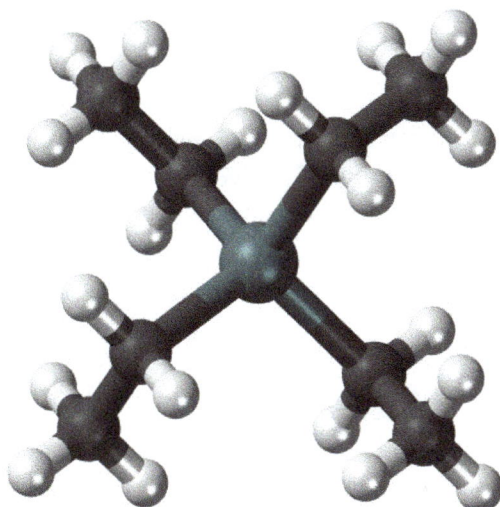

Tetraethyl lead is one of the most significant heavy metal contaminants in recent use.

Heavy metals are found naturally in the earth, and become concentrated as a result of human caused activities. Common sources are from mining and industrial wastes; vehicle emissions; lead-acid batteries; fertilisers; paints; treated woods; aging water supply infrastructure; and microplastics floating in the world's oceans. Arsenic, cadmium and lead may be present in children's toys at levels that exceed regulatory standards. Lead can be used in toys as a stabilizer, color enhancer, or anti-corrosive agent. Cadmium is sometimes employed as a stabilizer, or to increase the mass and luster of toy jewelry. Arsenic is thought to be used in connection with coloring dyes. Regular imbibers of illegally distilled alcohol may be exposed to arsenic or lead poisoning the source of which is arsenic-contaminated lead used to solder the distilling apparatus. Rat poison used in grain and mash stores may be another source of the arsenic.

Lead is the most prevalent heavy metal contaminant. As a component of tetraethyl lead, $(CH_3CH_2)_4Pb$, it was used extensively in gasoline during the 1930s–1970s. Lead levels in the aquatic environments of industrialised societies have been estimated to be two to three times those of pre-industrial levels. Although the use of leaded gasoline was largely phased out in North America by 1996, soils next to roads built before this time retain high lead concentrations. Lead (from lead azide or lead styphnate used in firearms) gradually accumulates at firearms training grounds, contaminating the local environment and exposing range employees to a risk of lead poisoning.

Entry Routes

Heavy metals enter plant, animal and human tissues via air inhalation, diet and manual handling. Motor vehicle emissions are a major source of airborne contaminants including arsenic, cadmium, cobalt, nickel, lead, antimony, vanadium, zinc, platinum, palladium and rhodium. Water sources (groundwater, lakes, streams and rivers) can be polluted by heavy metals leaching from industrial and consumer waste; acid rain can exacerbate this process by releasing heavy metals trapped in soils. Plants are exposed to heavy metals through the uptake of water; animals eat these plants; ingestion of plant- and animal-based foods are the largest sources of heavy metals in humans. Absorption through skin contact, for example from contact with soil, is another potential source of heavy metal contamination. Toxic heavy metals can bioaccumulate in organisms as they are hard to metabolize.

Detrimental Effects

Heavy metals "can bind to vital cellular components, such as structural proteins, enzymes, and nucleic acids, and interfere with their functioning." Symptoms and effects can vary according to the metal or metal compound, and the dose involved. Broadly, long-term exposure to toxic heavy metals can have carcinogenic, central and peripheral nervous system and circulatory effects. For humans, typical presentations associated with exposure to any of the "classical" toxic heavy metals, or chromium (another toxic heavy metal) or arsenic (a metalloid), are shown in the table.

Element	Acute exposure *usually a day or less*	Chronic exposure *often months or years*
Cadmium	Pneumonitis (lung inflammation)	Lung cancer Osteomalacia (softening of bones) Proteinuria (excess protein in urine; possible kidney damage)
Mercury	Diarrhea Fever Vomiting	Stomatitis (inflammation of gums and mouth) Nausea Nephrotic syndrome (nonspecific kidney disorder) Neurasthenia (neurotic disorder) Parageusia (metallic taste) Pink Disease (pain and pink discoloration of hands and feet) Tremor
Lead	Encephalopathy (brain dysfunction) Nausea Vomiting	Anemia Encephalopathy Foot drop/wrist drop (palsy) Nephropathy (kidney disease)
Chromium	Gastrointestinal hemorrhage (bleeding) Hemolysis (red blood cell destruction) Acute renal failure	Pulmonary fibrosis (lung scarring) Lung cancer

	Nausea Vomiting Diarrhea Encephalopathy Multi-organ effects Arrhythmia Painful neuropathy	Diabetes Hypopigmentation/Hyperkeratosis Cancer
Arsenic		

History

The toxic effects of arsenic, mercury and lead were known to the ancients but methodical studies of the overall toxicity of heavy metals appear to date from only 1868. In that year, Wanklyn and Chapman speculated on the adverse effects of the heavy metals "arsenic, lead, copper, zinc, iron and manganese" in drinking water. They noted an "absence of investigation" and were reduced to "the necessity of pleading for the collection of data." In 1884, Blake described an apparent connection between toxicity and the atomic weight of an element. The following sections provide historical thumbnails for the "classical" toxic heavy metals (arsenic, mercury and lead) and some more recent examples (chromium and cadmium).

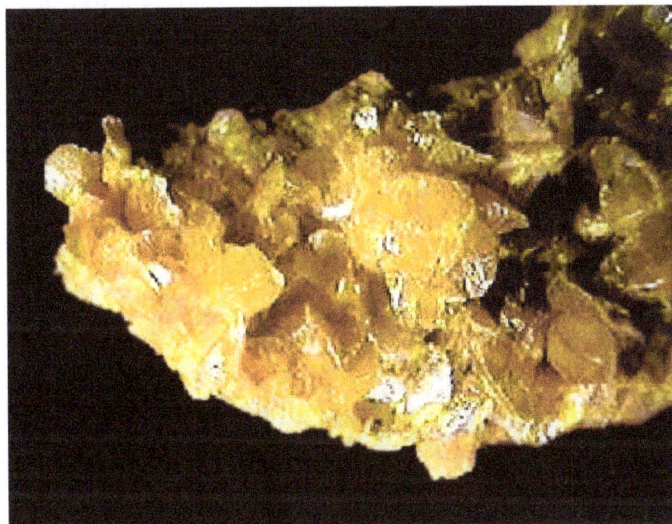

Orpiment, a toxic arsenic mineral used in the tanning industry to remove hair from hides.

Arsenic

Arsenic, as realgar (As_4S_4) and orpiment (As_2S_3), was known in ancient times. Strabo (64–50 BCE – c. AD 24?), a Greek geographer and historian, wrote that only slaves were employed in realgar and orpiment mines since they would inevitably die from the toxic effects of the fumes given off from the ores. Arsenic-contaminated beer poisoned over 6,000 people in the Manchester area of England in 1900, and is thought to have killed at least 70 victims. Clare Luce, American ambassador to Italy from 1953 to 1956, suffered from arsenic poisoning. Its source was traced to flaking arsenic-laden paint on the ceiling of her bedroom. She may also have eaten food contaminated by arsenic in flaking ceiling paint in the embassy dining room. Ground water contaminated by arsenic, as of 2014, "is still poisoning millions of people in Asia."

Mercury

Saint Isaac's Cathedral, in Saint Petersburg, Russia. The gold-mercury amalgam used to gild its dome caused numerous casualties among the workers involved.

The first emperor of unified China, Qin Shi Huang, it is reported, died of ingesting mercury pills that were intended to give him eternal life. The phrase "mad as a hatter" is likely a reference to mercury poisoning among milliners (so-called "mad hatter disease"), as mercury-based compounds were once used in the manufacture of felt hats in the 18th and 19th century. Historically, gold amalgam (an alloy with mercury) was widely used in gilding, leading to numerous casualties among the workers. It is estimated that during the construction of Saint Isaac's Cathedral alone, 60 workers died from the gilding of the main dome. Outbreaks of methylmercury poisoning occurred in several places in Japan during the 1950s due to industrial discharges of mercury into rivers and coastal waters. The best-known instances were in Minamata and Niigata. In Minamata alone, more than 600 people died due to what became known as Minamata disease. More than 21,000 people filed claims with the Japanese government, of which almost 3000 became certified as having the disease. In 22 documented cases, pregnant women who consumed contaminated fish showed mild or no symptoms but gave birth to infants with severe developmental disabilities. Since the industrial Revolution, mercury levels have tripled in many near-surface seawaters, especially around Iceland and Antarctica.

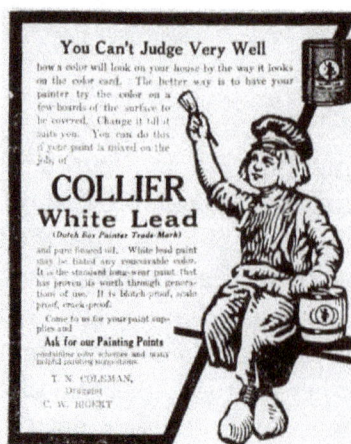

Dutch Boy white lead paint advertisement, 1912.

Lead

The adverse effects of lead were known to the ancients. In the 2nd century BC the Greek botanist Nicander described the colic and paralysis seen in lead-poisoned people. Dioscorides, a Greek physician who is thought to have lived in the 1st century CE, wrote that lead "makes the mind give way". Lead was used extensively in Roman aqueducts from about 500 BC to 300 AD. Julius Caesar's engineer, Vitruvius, reported, "water is much more wholesome from earthenware pipes than from lead pipes. For it seems to be made injurious by lead, because white lead is produced by it, and this is said to be harmful to the human body." During the Mongol period in China (1271–1368 AD), lead pollution due to silver smelting in the Yunnan region exceeded contamination levels from modern mining activities by nearly four times.[n 1] In the 17th and 18th centuries, people in Devon were afflicted by a condition referred to as Devon colic; this was discovered to be due to the imbibing of lead-contaminated cider. In 2013, the World Health Organization estimated that lead poisoning resulted in 143,000 deaths, and "contribute[d] to 600,000 new cases of children with intellectual disabilities", each year. In north-east America, in the city of Flint, Michigan, lead contamination in drinking water has been an issue since 2014. The source of the contamination has been attributed to "corrosion in the lead and iron pipes that distribute water to city residents". In 2015, drinking water lead levels in north-eastern Tasmania, Australia, were reported to reach over 50 times national drinking water guidelines. The source of the contamination was attributed to "a combination of dilapidated drinking water infrastructure, including lead jointed pipelines, end-of-life polyvinyl chloride pipes and household plumbing."

Chromium

Potassium chromate, a carcinogen, is used in the dyeing of fabrics, and as a tanning agent to produce leather.

Chromium(III) compounds and chromium metal are not considered a health hazard, while the toxicity and carcinogenic properties of chromium(VI) have been known since at least the late 19th century. In 1890, Newman described the elevated cancer risk of workers in a chromate dye company. Chromate-induced dermatitis was reported in aircraft workers during World War II. In 1963, an outbreak of dermatitis, ranging from erythema to exudative eczema, occurred amongst 60 automobile factory workers in England. The workers had been wet-sanding chromate-based primer paint that had been applied to car bodies. In Australia, chromium was released from the Newcastle Orica explosives plant on August 8, 2011. Up to 20 workers at the plant were exposed as were 70 nearby homes in Stockton. The town was only notified three days after the release and the accident sparked a major public controversy, with Orica criticised for playing down the extent and possible risks of the leak, and the state Government attacked for their slow response to the incident.

99.999% purity cadmium bar and 1 cm^3 cube.

Cadmium

Cadmium exposure is a phenomenon of the early 20th century, and onwards. In Japan in 1910, the Mitsui Mining and Smelting Company began discharging cadmium into the Jinzugawa river, as a byproduct of mining operations. Residents in the surrounding area subsequently consumed rice grown in cadmium-contaminated irrigation water. They experienced softening of the bones and kidney failure. The origin of these symptoms was not clear; possibilities raised at the time included "a regional or bacterial disease or lead poisoning." In 1955, cadmium was identified as the likely cause and in 1961 the source was directly linked to mining operations in the area. In February 2010, cadmium was found in Wal-Mart exclusive Miley Cyrus jewelry. Wal-Mart continued to sell the jewelry until May, when covert testing organised by Associated Press confirmed the original results. In June 2010 cadmium was detected in the paint used on promotional drinking glasses for the movie Shrek Forever After, sold by McDonald's Restaurants, triggering a recall of 12 million glasses.

Remediation

A metal EDTA anion. Pb displaces Ca in $Na_2[CaEDTA]$ to give $Na_2[PbEDTA]$, which is passed out of the body in urine.

In humans, heavy metal poisoning is generally treated by the administration of chelating agents. These are chemical compounds, such as CaNa2 EDTA (calcium disodium ethylenediaminetetraacetate) that convert heavy metals to chemically inert forms that can be excreted without further interaction with the body. Chelates are not without side effects and can also remove beneficial metals from the body. Vitamin and mineral supplements are sometimes co-administered for this reason.

Soils contaminated by heavy metals can be remediated by one or more of the following technologies: isolation; immobilization; toxicity reduction; physical separation; or extraction. *Isolation* involves the use of caps, membranes or below-ground barriers in an attempt to quarantine the contaminated soil. *Immobilization* aims to alter the properties of the soil so as to hinder the mobility of the heavy contaminants. *Toxicity reduction* attempts to oxidise or reduce the toxic heavy metal ions, via chemical or biological means into less toxic or mobile forms. *Physical separation* involves the removal of the contaminated soil and the separation of the metal contaminants by mechanical means. *Extraction* is an on or off-site process that uses chemicals, high-temperature volatization, or electrolysis to extract contaminants from soils. The process or processes used will vary according to contaminant and the characteristics of the site.

Benefits

Some elements otherwise regarded as toxic heavy metals are essential, in small quantities, for human health. These elements include vanadium, manganese, iron, cobalt, copper, zinc, selenium, strontium and molybdenum. A deficiency of these essential metals may increase susceptibility to heavy metal poisoning.

Arsenic Contamination of Groundwater

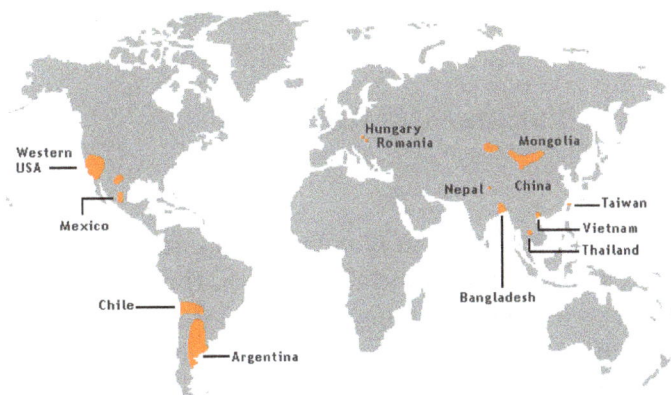

Groundwater arsenic contamination areas

Arsenic contamination of groundwater is a form of groundwater pollution which is often due to naturally occurring high concentrations of arsenic in deeper levels of groundwater. It is a high-profile problem due to the use of deep tubewells for water supply in the Ganges Delta, causing serious arsenic poisoning to large numbers of people. A 2007 study found that over 137 million people in more than 70 countries are probably affected by arsenic poisoning of drinking water. Arsenic contamination of ground water is found in many countries throughout the world, including the USA.

Approximately 20 major incidents of groundwater arsenic contamination have been reported. Of these, four major incidents occurred in Asia, in Thailand, Taiwan, and Mainland China.

Speciation of Arsenic Compounds in Water

Arsenic contaminated water typically contains arsenous acid and arsenic acid or their derivatives. Their names as "acids" is a formality, these species are not aggressive acids but are merely the soluble forms of arsenic near neutral pH. These compounds are extracted from the underlying rocks that surround the aquifer. Arsenic acid tends to exist as the ions $[HAsO_4]^{2-}$ and $[H_2AsO_4]^-$ in neutral water, whereas arsenous acid is not ionized.

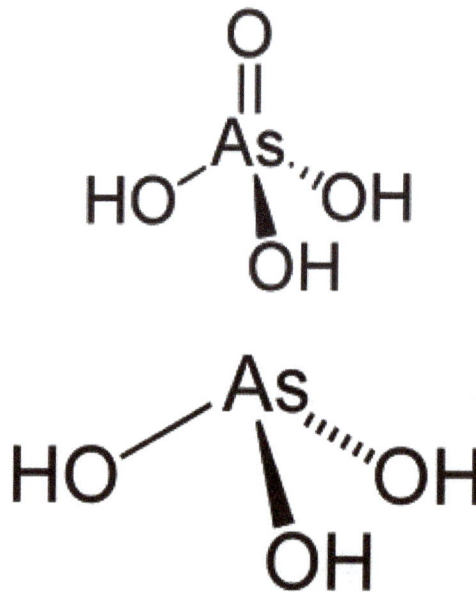

Arsenic acid (H_3AsO_4), arsenous acid (H_3AsO_3) and their derivatives are typically encountered in arsenic contaminated ground water.

Contamination Specific Nations and Regions

India and Bangladesh

Arsenic contamination of the groundwater in Bangladesh is a serious problem. Prior to the 1970s, Bangladesh had one of the highest infant mortality rates in the world. Ineffective water purification and sewage systems as well as periodic monsoons and flooding exacerbated these problems. As a solution, UNICEF and the World Bank advocated the use of wells to tap into deeper groundwater. Millions of wells were constructed as a result. Because of this action, infant mortality and diarrheal illness were reduced by fifty percent. However, with over 8 million wells constructed, approximately one in five of these wells is now contaminated with arsenic above the government's drinking water standard.

In the Ganges Delta, the affected wells are typically more than 20 meters and less than 100 meters deep. Groundwater closer to the surface typically has spent a shorter time in the ground, therefore likely absorbing a lower concentration of arsenic; water deeper than 100 m is exposed to much older sediments which have already been depleted of arsenic.

The issue came to international attention in 1995. The study conducted in Bangladesh involved the analysis of thousands of water samples as well as hair, nail, and urine samples. They found 900 villages with arsenic above the government limit.

Criticism has been leveled at the aid agencies, who denied the problem during the 1990s while millions of tube wells were sunk. The aid agencies later hired foreign experts who recommended treatment plants that were inappropriate to the conditions, were regularly breaking down, or were not removing the arsenic.

In West Bengal, India, water is mostly supplied from rivers. Groundwater comes from deep tubewells, which are few in number. Because of the low quantity of deep tubewells, the risk of arsenic poisoning in West Bengal is comparatively less. According to the World Health Organisation, "In Bangladesh, West Bengal (India), and some other areas most drinking-water used to be collected from open dug wells and ponds with little or no arsenic, but with contaminated water transmitting diseases such as diarrhoea, dysentery, typhoid, cholera, and hepatitis. Programmes to provide 'safe' drinking-water over the past 30 years have helped to control these diseases, but in some areas they have had the un-expected side-effect of exposing the population to another health problem—arsenic." The acceptable level as defined by WHO for maximum concentrations of arsenic in safe drinking water is 0.01 mg/L. The Bangladesh government's standard is a fivefold greater rate, with 0.05 mg/L being considered safe. WHO has defined the areas under threat: Seven of the nineteen districts of West Bengal have been reported to have ground water arsenic concentrations above 0.05 mg/L. The total population in these seven districts is over 34 million while the number using arsenic-rich water is more than 1 million (above 0.05 mg/L). That number increases to 1.3 million when the concentration is above 0.01 mg/L. According to a British Geological Survey study in 1998 on shallow tube-wells in 61 of the 64 districts in Bangladesh, 46 percent of the samples were above 0.01 mg/L and 27 percent were above 0.050 mg/L. When combined with the estimated 1999 population, it was estimated that the number of people exposed to arsenic concentrations above 0.05 mg/L is 28-35 million and the number of those exposed to more than 0.01 mg/L is 46-57 million (BGS, 2000).

Throughout Bangladesh, as tube wells get tested for concentrations of arsenic, ones which are found to have arsenic concentrations over the amount considered safe are painted red to warn residents that the water is not safe to drink.

One solution is "By using surface water and instituting effective withdrawal regulation. West Bengal and Bangladesh are flooded with surface water. We should first regulate proper watershed management. Treat and use available surface water, rain-water, and others. The way we're doing [it] at present is not advisable.". Another avenue would be looking at the nutrition content which is also seen to be responsible for mitigating the effects of Arsenic. Malnutrition is seen to increase the adverse effects of arsenic.

In Bihar, Groundwater in 13 districts have been found to be contaminated with Arsenic with quantities exceeding 0.05 mg/L. All these districts are situated close to large rivers like Ganga and Gandak.

Argentina

The central portion of Argentina is affected by arsenic-contaminated groundwater. Specifically, the La Pampa produces water containing 4-5300 microgram As per litre.

United States

Regulation

A drinking water standard of 0.05 mg/L (equal to 50 parts per billion, or ppb) arsenic was originally established in the United States by the Public Health Service in 1942. The Environmental Protection Agency (EPA) studied the pros and cons of lowering the arsenic Maximum Contaminant Level (MCL) for years in the late 1980s and 1990s. No action was taken until January 2001, when the Clinton administration in its final weeks promulgated a new standard of 0.01 mg/L (10 ppb) to take effect January 2006. The Bush administration suspended the midnight regulation, but after some months of study, the new EPA administrator Christine Todd Whitman approved the new 10 ppb arsenic standard and its original effective date of January 2006. Many locations exceed this limit. Cases of groundwater-caused acute arsenic toxicity, such as those found in Bangladesh, are unknown in the United States where the concern has focused on the role of arsenic as a carcinogen. The problem of high arsenic concentrations has been subject to greater scrutiny since 2001, when the federal government promulgated a new lower standard for arsenic in drinking water.

Many public water supply systems across the United States obtained their water supply from groundwater that had met the old 50 ppb arsenic standard but exceeded the new 10 ppb MCL. These utilities searched for either an alternative supply or an inexpensive treatment method to remove the arsenic from their water. In Arizona, an estimated 35 percent of water-supply wells were put out of compliance by the new regulation; in California, the percentage was 38 percent .

The proper arsenic MCL continues to be debated. Some have argued that the 10 ppb federal standard is still too high, while others have argued that 10 ppb is needlessly strict. Individual states are able to establish lower arsenic limits; New Jersey has done so, setting a maximum of 0.005 mg/L for arsenic in drinking water.

A study of private water wells in the Appalachian mountains found that six percent of the wells had arsenic above the U.S. MCL of 0.010 mg/L.

Case Studies and Incidents

Fallon, Nevada has long been known to have groundwater with relatively high arsenic concentrations (in excess of 0.08 mg/L). Even some surface waters, such as the Verin Arizona, sometimes exceed 0.01 mg/L arsenic, especially during low-flow periods when the river flow is dominated by groundwater discharge.

A study conducted in a contiguous six-county area of southeastern Michigan investigated the relationship between moderate arsenic levels and 23 selected disease outcomes. Disease outcomes included several types of cancer, diseases of the circulatory and respiratory system, diabetes mellitus, and kidney and liver diseases. Elevated mortality rates were observed for all diseases of the circulatory system. The researchers acknowledged a need to replicate their findings.

Nepal

Nepal is subject to a serious problem with arsenic contamination. The problem is most severe in the Terai region, the worst being near Nawalparasi District, where 26 percent of shallow wells failed to meet WHO standard of 10 ppb. A study by Japan International Cooperation Agency and the Environment in the Kathmandu Valley showed that 72% of deep wells failed to meet the WHO standard, and 12% failed to meet the Nepali standard of 50 ppb.

Water Purification Solutions

Small-Scale Water Treatment

A review of methods to remove arsenic from groundwater in Pakistan summarizes the most technically viable inexpensive methods.

A simpler and less expensive form of arsenic removal is known as the Sono arsenic filter, using three pitchers containing cast iron turnings and sand in the first pitcher and wood activated carbon and sand in the second. Plastic buckets can also be used as filter containers. It is claimed that thousands of these systems are in use and can last for years while avoiding the toxic waste disposal problem inherent to conventional arsenic removal plants. Although novel, this filter has not been certified by any sanitary standards such as NSF, ANSI, WQA and does not avoid toxic waste disposal similar to any other iron removal process.

In the United States small "under the sink" units have been used to remove arsenic from drinking water. This option is called "point of use" treatment. The most common types of domestic treatment use the technologies of adsorption (using media such as Bayoxide E33, GFH, or titanium dioxide) or reverse osmosis. Ion exchange and activated alumina have been considered but not commonly used.

Chaff-based filters have been reported to reduce the arsenic content of water to 3 microgram/litre. This is especially important in areas where the potable water is provided by filtering the water extracted from the underground aquifer.

Large-scale Water Treatment

In some places, such as the United States, all the water supplied to residences by utilities must meet primary (health-based) drinking water standards. Regulations may necessitate large-scale treatment systems to remove arsenic from the water supply. The effectiveness of any method depends on the chemical makeup of a particular water supply. The aqueous chemistry of arsenic is complex, and may affect the removal rate that can be achieved by a particular process.

Some large utilities with multiple water supply wells could shut down those wells with high arsenic concentrations, and produce only from wells or surface water sources that meet the arsenic standard. Other utilities, however, especially small utilities with only a few wells, may have no available water supply that meets the arsenic standard.

Coagulation/filtration (also known as flocculation) removes arsenic by coprecipitation and adsorption using iron coagulants. Coagulation/filtration using alum is already used by some utilities to remove suspended solids and may be adjusted to remove arsenic. But the problem of this type of filtration system is that it gets clogged very easily, mostly within two to three months. The toxic arsenic sludge are disposed of by concrete stabilization, but there is no guarantee that they won't leach out in future.

Iron oxide adsorption filters the water through a granular medium containing ferric oxide. Ferric oxide has a high affinity for adsorbing dissolved metals such as arsenic. The iron oxide medium eventually becomes saturated, and must be replaced. The sludge disposal is a problem here too.

Activated alumina is an adsorbent that effectively removes arsenic. Activated alumina columns connected to shallow tube wells in India and Bangladesh have removed both As(III) and As(V) from groundwater for decades. Long-term column performance has been possible through the efforts of community-elected water committees that collect a local water tax for funding operations and maintenance. It has also been used to remove undesirably high concentrations of fluoride.

Ion exchange has long been used as a water softening process, although usually on a single-home basis. Traditional anion exchange resins are effective in removing As(V), but not As(III), or arsenic trioxide, which doesn't have a net charge. Effective long-term ion exchange removal of arsenic requires a trained operator to maintain the column.

Both Reverse osmosis and electrodialysis (also called *electrodialysis reversal*) can remove arsenic with a net ionic charge. (Note that arsenic oxide, As_2O_3, is a common form of arsenic in groundwater that is soluble, but has no net charge.) Some utilities presently use one of these methods to reduce total dissolved solids and therefore improve taste. A problem with both methods is the production of high-salinity waste water, called brine, or concentrate, which then must be disposed of.

Subterranean Arsenic Removal (SAR) Technology

In subterranean arsenic removal (SAR), aerated groundwater is recharged back into the aquifer to create an oxidation zone which can trap iron and arsenic on the soil particles through adsorption process. The oxidation zone created by aerated water boosts the activity of the arsenic-oxidizing microorganisms which can oxidize arsenic from +3 to +5 state SAR Technology. No chemicals are used and almost no sludge is produced during operational stage since iron and arsenic compounds are rendered inactive in the aquifer itself. Thus toxic waste disposal and the risk of its future mobilization is prevented. Also, it has very long operational life, similar to the long lasting tube wells drawing water from the shallow aquifers.

Six such SAR plants, funded by the World Bank and constructed by Ramakrishna Vivekananda Mission, Barrackpore & Queen's University Belfast, UK are operating in West Bengal. Each plant has been delivering more than 3,000 litres of arsenic and iron-free water daily to the rural community. The first community water treatment plant based on SAR technology was set up at Kashimpore near Kolkata in 2004 by a team of European and Indian engineers led by Bhaskar Sen Gupta of Queen's University Belfast for TiPOT.

SAR technology had been awarded Dhirubhai Ambani Award, 2010 from IChemE UK for Chemical Innovation. Again, SAR was the winner of the St. Andrews Award for Environment, 2010. The SAR Project was selected by the Blacksmith Institute - New York & Green Cross- Switzerland as one of the "12 Cases of Cleanup & Success" in the World's Worst Polluted Places Report 2009. (Refer: www.worstpolluted.org).

Currently, large scale SAR plants are being installed in USA, Malaysia, Cambodia, and Vietnam.

Dietary Intake

Researchers from Bangladesh and the United Kingdom have claimed that dietary intake of arsenic adds a significant amount to total intake where contaminated water is used for irrigation.

Mercury in Fish

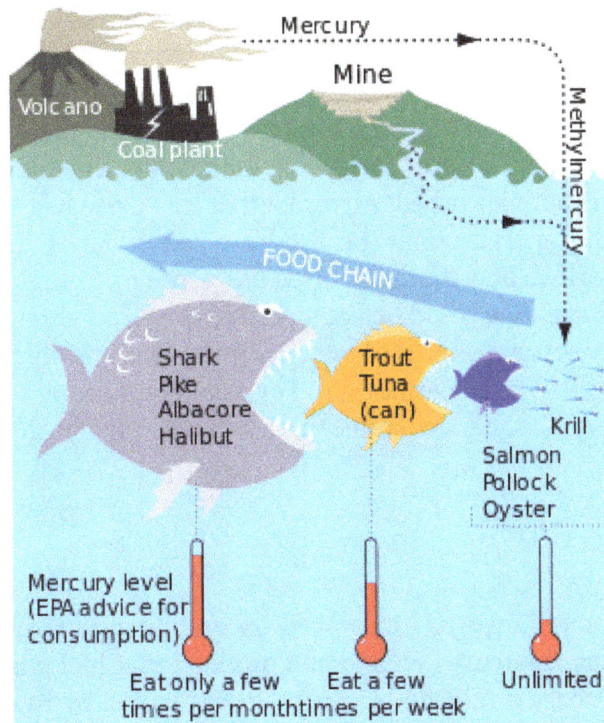

Nearby anthropogenic sources, such as coal burning and mining of iron, can contaminate water sources with methylmercury, which is efficiently absorbed in the bodies of fish. Through the process of biomagnification, mercury levels in each successive predatory stage increase.

Fish and shellfish concentrate mercury in their bodies, often in the form of methylmercury, a highly toxic organic compound of mercury. Fish products have been shown to contain varying amounts of heavy metals, particularly mercury and fat-soluble pollutants from water pollution. Species of fish that are long-lived and high on the food chain, such as marlin, tuna, shark, swordfish, king mackerel, tilefish (Gulf of Mexico), and northern pike, contain higher concentrations of mercury than others.

Mercury is known to bioaccumulate in humans, so bioaccumulation in seafood carries over into human populations, where it can result in mercury poisoning. Mercury is dangerous to both natural ecosystems and humans because it is a metal known to be highly toxic, especially due to its ability to damage the central nervous system. In human-controlled ecosystems of fish, usually done for market production of wanted seafood species, mercury clearly rises through the food chain via fish consuming small plankton, as well as through non-food sources such as underwater sediment. This mercury grows in concentration within the bodies of fish and can be measured in the tissues of selected species.

The presence of mercury in fish can be a particular health concern for women who are or may become pregnant, nursing mothers, and young children.

Biomagnification

The consumption of fish is by far the most significant source of ingestion-related mercury exposure in humans and animals. Mercury and methyl mercury are present in only very small concentrations in seawater. However, they are absorbed, usually as methyl mercury, by algae at the start of the food chain. This algae is then eaten by fish and other organisms higher in the food chain. Fish efficiently absorb methyl mercury, but only very slowly excrete it. Methyl mercury is not soluble and therefore is not apt to be excreted. Instead, it accumulates, primarily in the viscera, although also in the muscle tissue. This results in the bioaccumulation of mercury, in a buildup in the adipose tissue of successive trophic levels: zooplankton, small nekton, larger fish, and so on. The older that such fish become, the more mercury they may have absorbed. Anything that eats these fish within the food chain also consumes the higher level of mercury that the fish have accumulated. This process explains why predatory fish such as swordfish and sharks or birds like osprey and eagles have higher concentrations of mercury in their tissue than could be accounted for by direct exposure alone. Species on the food chain can amass body concentrations of mercury up to ten times higher than the species they consume. This process is called biomagnification. For example, herring contains mercury levels at about 0.1 parts per million, while shark contains mercury levels greater than 1 part per million.

Legislation

Japan

Since the Minamata disaster, Japan has improved on its mercury regulation. During the 1970s Japan made strides to reduce mercury demand and production. Chief among these efforts was the reduction of inorganic mercury produced by mines. It was halted by 1974, and demand fell from 2,500 tons per year in 1964, its peak, to 10 tons per year in recent years. Since these initial strides, Japan has introduced a list of regulations governing the mercury content of a variety of materials.

Japan Mercury Regulation		
Category	**Regulation**	**Result**
Cosmetics	Pharmaceutical Affairs Act	Ban the use of mercury and its compounds
Agriculture	Agricultural Chemicals Control Act	Ban the use of mercury and its compounds as an active ingredient
Household Commodities	No mercury in household adhesives, household paints, household wax, shoe polish, shoe cream, diapers, bibs, undergarments, gloves, and socks	Act on Control of Household Products Containing Hazardous Substances
Pharmaceutical Products	No use of mercury compounds in oral preparations. No use of mercury compounds, other than mercurochrome, as an active ingredient. Mercury as a preservative only if no other option is available.	Pharmaceutical Affairs Act
Air	No more than 40 ng/m³	Air Pollution Control Law

Water	Environmental quality standard: no more than 0.0005 mg/L in waterway and ground water. Effluent standard: no more than 0.005 mg/L in effluence.	Basic Environment Law and Water Pollution Control Act
Soil	Environmental quality standard: no more than 0.0005 mg/L sample solution. Elution standard: no more than 0.0005 mg/L. Content standard: no more than 15 mg/kg	Basic Environment Law and Soil Contamination Countermeasures Act

Regulation of these potential sources of pollution reduces the amount of mercury that ends up in fish and, through biomagnification, in humans. In addition to enacting legislation controlling the mercury levels in potential pollutants, Japan has directly influenced the environment by issuing regulations setting acceptable levels of environmental mercury pollution.

It is Japan's goal to promote international mercury legislation in hopes of preventing any country from experiencing what it did. Despite Japan's extensive regulation and experience with mercury-based disasters, there is still little information provided to the public. The Japanese Federal Fish Advisory's recommendations are less strict than those in America.

United States of America

The United States is a leader in mercury regulation. A key piece of mercury legislation in the United States is the *Mercury and Air Toxics Standards* (MATS). This policy was finalized by the Environmental Protection Agency (EPA) on December 16, 2011. This is a federal policy which directly influences mercury in fish, and is the first of its kind in the United States. The facilities targeted by this new policy are the chief sources of mercury in the air. The airborne mercury is dissolved in the oceans, where microorganisms convert waterborne mercury into methyl mercury; mercury thus finds its way into the food chain and into fish. MATS is legislated towards the aim of preventing about 90% of the emissions from power plants from reaching the air. In total the expected health benefits are estimated at $37 billion–$90 billion by 2016. In comparison, the expected economic cost is $9.6 billion annually. Another integral piece of legislation controlling the emission of mercury to the air is the Clean Air Act. Under this act, mercury is classified as a hazardous air pollutant, allowing the EPA to regulate emissions by establishing performance standards.

International

Legislation on a global scale is believed by some to be needed for this issue because mercury pollution is estimated to be so far-reaching. Pollution from one country does not stay localized to that country. Despite the need by some, international regulation has been slow to take off. The first forms of international legislation appeared in the 1970s, beginning as agreements about shared bodies of water. The next step was the Stockholm Declaration, which urged countries to avoid polluting the oceans by dumping. The 1972 Oslo Convention and the 1974 Paris Convention were adopted by parts of Europe. Both lessened polluting the ocean with mercury, the former by banning the dumping of ships and aircraft into the ocean and the latter by obligating participants to reduce land-based pollution on coastlines. The first real global legislation regarding mercury pollution was the Basel Convention of 1989. This convention attempts to reduce the movement of mercury across borders and primarily regulates the import and export of toxic chemicals, including mercury. In 1998 the Convention on Long-Range Transboundary Air Pollution was adopted by most of the European Union, the United States, and Canada. Its primary objective is to cut emissions of heavy metals. The convention is the largest international agreement on mercury established to date. In the early 21st century, the focus of mercury regulation has been on voluntary programs. The next phase in legislation is a global effort, and this appears to be what the Minamata Convention hopes to accomplish. The Minamata Convention, named after the Japanese city that suffered horribly from mercury pollution, has taken four years of negotiation but was finally adopted by delegates from over 140 countries. The convention will come into power after 50 countries have signed it. The Minamata Convention will require all participants to eliminate, where possible, the release of mercury from small-scale gold mining. It will also require a sharp reduction in emission from coal burning.

Levels of Contamination

Most-contaminated Fish Species

The danger level from consuming fish depends on species and size. Size is the best predictor of increased levels of accumulated mercury. Sharks, such as the mako shark, have very high levels of mercury. A study on New Jersey coastal fish indicated that one third of the sampled fish had levels of mercury above 0.5 parts per million, a level that could pose a human health concern for consumers who regularly eat this fish. Another study of marketplace fish caught in waters surrounding Southern Italy showed that, undoubtedly, greater fish weight leads to additional mercury found in fish body tissues. Moreover, the concentration, measured in milligrams of mercury per kilogram of fish, steadily increases with the size of the fish. Anglerfish off the coast of Italy were found with concentrations as high as 2.2 milligrams of mercury per kilogram, higher than the recommended limit of 1 milligram of mercury per kilogram. Annually, Italy catches approximately a third of its fish from the Adriatic Sea, where these anglerfish were found.

Fish that consume their prey in a certain manner may contain much higher concentrations of mercury than other species. Grass carp off the coast of China hold far less internal mercury than do bighead carp. The reason for this is that bighead carp are filter feeders, while grass carp are not. Thus, bighead carp gather more mercury by eating large amounts of small plankton, as well as sucking up sediments that collect a sizable amount of methyl mercury.

Mercury levels in commercial fish and shellfish						
species	Mean (ppm)	Std dev (ppm)	Median (ppm)	Comment	Trophic level	Max age (years)
Tilefish (Gulf of Mexico)	1.450	n/a	n/a	Mid-Atlantic tilefish has lower mercury levels and is considered safe to eat in moderation.	3.6	35
Swordfish	0.995	0.539	0.870		4.5	15
Shark	0.979	0.626	0.811			
Mackerel (king)	0.730	n/a	n/a		4.5	14

Tuna (bigeye)	0.689	0.341	0.560	Fresh/frozen	4.5	11
Orange roughy	0.571	0.183	0.562		4.3	149
Marlin *	0.485	0.237	0.390		4.5	
Mackerel (Spanish)	0.454	n/a	n/a	Gulf of Mexico	4.5	5
Grouper	0.448	0.278	0.399	All species	4.2	
Tuna	0.391	0.266	0.340	All species, fresh/frozen		
Bluefish	0.368	0.221	0.305		4.5	9
Sablefish	0.361	0.241	0.265		3.8	114
Tuna (albacore)	0.358	0.138	0.360	Fresh/frozen	4.3	9
Patagonian toothfish	0.354	0.299	0.303	AKA Chilean sea bass	4.0	50+
Tuna (yellowfin)	0.354	0.231	0.311	Fresh/frozen	4.3	9
Tuna (albacore)	0.350	0.128	0.338	Canned	4.3	9
Croaker white	0.287	0.069	0.280	Pacific	3.4	
Halibut	0.241	0.225	0.188		4.3	
Weakfish	0.235	0.216	0.157	Sea trout	3.8	17
Scorpionfish	0.233	0.139	0.181			
Mackerel (Spanish)	0.182	n/a	n/a	South Atlantic	4.5	
Monkfish	0.181	0.075	0.139		4.5	25
Snapper	0.166	0.244	0.113			
Bass	0.152	0.201	0.084	Striped, black, and black sea	3.9	
Perch	0.150	0.112	0.146	Freshwater	4.0	
Tilefish (Atlantic)	0.144	0.122	0.099		3.6	35
Tuna (skipjack)	0.144	0.119	0.150	Fresh/frozen	3.8	12
Buffalofish	0.137	0.094	0.120			
Skate	0.137	n/a	n/a			
Tuna	0.128	0.135	0.078	All species, canned, light		
Perch (ocean) *	0.121	0.125	0.102			
Cod	0.111	0.152	0.066		3.9	22
Carp	0.110	0.099	0.134			
Lobster (American)	0.107	0.076	0.086			
Sheephead (California)	0.093	0.059	0.088			
Lobster (spiny)	0.093	0.097	0.062			
Whitefish	0.089	0.084	0.067			
Mackerel (chub)	0.088	n/a	n/a	Pacific	3.1	

Herring	0.084	0.128	0.048		3.2	21
Jacksmelt	0.081	0.103	0.050		3.1	
Hake	0.079	0.064	0.067		4.0	
Trout	0.071	0.141	0.025	Freshwater		
Crab	0.065	0.096	0.050	Blue, king and snow crab		
Butterfish	0.058	n/a	n/a		3.5	
Flatfish *	0.056	0.045	0.050	Flounder, plaice and sole		
Haddock	0.055	0.033	0.049	Atlantic		
Whiting	0.051	0.030	0.052			
Mackerel (Atlantic)	0.050	n/a	n/a			
Croaker (Atlantic)	0.065	0.050	0.061			
Mullet	0.050	0.078	0.014			
Shad (American)	0.039	0.045	0.045			
Crayfish	0.035	0.033	0.012			
Pollock	0.031	0.089	0.003			
Catfish	0.025	0.057	0.005		3.9	24
Squid	0.023	0.022	0.016			
Salmon *	0.022	0.034	0.015	Fresh/frozen		
Anchovies	0.017	0.015	0.014		3.1	
Sardine	0.013	0.015	0.010		2.7	
Tilapia *	0.013	0.023	0.004			
Oyster	0.012	0.035	n/d			
Clam *	0.009	0.011	0.002			
Salmon *	0.008	0.017	n/d	Canned		
Scallop	0.003	0.007	n/d			
Shrimp *	0.001	0.013	0.009			6.5

* indicates only methylmercury was analyzed (all other results are for total mercury)
n/a – data not available
n/d – below detection level (0.01ppm)

US government scientists tested fish in 291 streams around the country for mercury contamination. They found mercury in every fish tested, according to the study by the U.S. Department of the Interior. They found mercury even in fish of isolated rural waterways. Twenty-five percent of the fish tested had mercury levels above the safety levels determined by the U.S. Environmental Protection Agency for people who eat the fish regularly.

Origins of Mercury Pollution

There are three types of mercury emission: anthropogenic, re-emission, and natural, including volcanoes and geothermal vents. Anthropogenic sources are responsible for 30% of all emissions, while natural sources are responsible for 10%, and re-emission accounts for the other 60%. While re-emission accounts for the largest proportion of emissions, it is likely that the mercury emitted from these sources originally came from anthropogenic sources.

Anthropogenic sources include coal burning, cement production, oil refining, artisan and small-scale gold mining, wastes from consumer products, dental amalgam, the chlor-alkali industry, production of vinyl chloride, and the mining, smelting, and production of iron and other metals. The total amount of mercury released by mankind in 2010 was estimated to be 1960 metric tons. The majority of this comes from coal burning and gold mining, accounting for 24% and 37% of total anthropogenic output respectively.

Re-emission, the largest emitter, occurs in a variety of ways. It is possible for mercury that has been deposited in soil to be re-emitted into the mercury cycle via floods. A second example of re-emission is a forest fire; mercury that has been absorbed into plant life is re-released into the atmosphere. While it is difficult to estimate the exact extent of mercury re-emission, it is an important field of study. Knowing how easily and how often previously emitted mercury can be released helps us learn how long it will take for a reduction in anthropogenic sources to be reflected in the environment. Mercury that has been released can find its way into the oceans. A 2008 model estimated the total amount of deposition into the oceans that year to be 3,700 metric tons. It is estimated that rivers carry as much as 2,420 metric tons. Much of the mercury deposited in the oceans is re-emitted, however; as much as 300 metric tons is converted into methyl mercury. While only 13% of this finds its way into the food chain, that is still 40 metric tons a year.

Much (an estimated 40%) of the mercury that eventually finds its way into fish originates with coal-burning power plants and chlorine production plants. The largest source of mercury contamination in the United States is coal-fueled power plant emissions. Chlorine chemical plants use mercury to extract chlorine from salt, which in many parts of the world is discharged as mercury compounds in waste water, though this process has been largely replaced by the more economically viable membrane cell process, which does not use mercury. Coal contains mercury as a natural contaminant. When it is fired for electricity generation, the mercury is released as smoke into the atmosphere. Most of this mercury pollution can be eliminated if pollution-control devices are installed.

Mercury in the United States frequently comes from power plants, which release about 50% of the nation's mercury emissions. In other countries, such as Ghana, gold mining requires mercury compounds, leading to workers receiving significant quantities of mercury while performing their jobs. Such mercury from gold mines is specifically known to contribute to biomagnification in aquatic food chains.

The farming of aquatic organisms, known as aquaculture, often involves fish feed that contains mercury. A study by Jardine has found no reliable connection between mercury in fish food affecting aquaculture organisms or aquatic organisms in the wild. Even so, mercury from other sources may still affect organisms grown through aquaculture. In China, farmed fish species, such as bighead carp, mud carp, and mandarin fish, carried 90% of total mercury content in all of the measured fish in a study by Cheng. This study also concluded that mercury bioaccumulates through food chains even in controlled aquaculture environments. Both total mercury and methyl mercury absorption was found to be derived from sediments containing mercury, not mainly from fish feed.

The Hawaii Institute of Marine Biology has noted that fish feed used in aquaculture often contains heavy metals such as mercury, lead, and arsenic, and has dispatched these concerns to organizations such as the Food and Agriculture Organization of the United Nations. An industry aquaculture company, Aquatic Farms Ltd., has posted numerous guidelines on how to keep fish feed safe, thus protecting the consumer from any dangers posed by the food eaten by commercial seafood species. Some guidelines include providing a cool and dry environment to prevent both spoilage and contamination of the fish feed and avoiding pesticides and other chemicals up to a certain limit.

Elemental mercury often comes from coal power plants, and oxidized mercury often comes from incinerators. Oil-fired power plants also contribute mercury to the environment. The energy industry therefore is a key player in the introduction of mercury into the environment. When addressing the issue of reducing seafood mercury bioaccumulation on a global scale, it is important to pinpoint major energy producers and consumers whose exchange of energy may be the root of the problem.

Controlling Output of Mercury Pollution Sources

A study that was led by scientists from Harvard University and U.S. Geological Survey has determined that in the next several decades there will be a 50 percent increase in mercury levels. The study also shows that the increases are connected through industrial emissions and are not natural as previously thought. However, by decreasing emissions from industrial plants, the possibility of decreasing the high level of mercury remains plausible. Several nations are currently implementing systems that will detect and therefore later be able to control the output of mercury into the atmosphere. Air pollution control devices (APCDs) have been implemented in South Korea as the government is starting to take inventory of mercury sources. Mercury pollution can also be removed by electrostatic precipitators (ESPs). Bag-based filters are also used in factories that may contribute mercury to the environment. Flue-gas desulfurization, normally used to eliminate sulfur dioxide, can also be used in conjunction with APCDs to remove additional mercury before exhausts are released into the environment. Even so, countries such as South Korea have only begun to use inventories of mercury sources, calling into question how fast anti-mercury measures will be put into factories.

Health Effects and Outcomes

Disparate Impacts

Mercury content in fish does not affect all populations equally. Certain ethnic groups, as well as young children, are more likely to suffer the effects of methyl mercury poisoning. In the United States, Wallace gathered data that indicated 16.9% of women who self-identify as Native American, Asian, Pacific Islander, or multiracial exceed the recommended reference dose of mercury. A study done on children of the Faroe Islands near Great Britain showed neurological problems stemming from mothers consuming pilot whale meat during pregnancy. Such data demonstrate that certain ethnic groups, as well as children, are particularly vulnerable to methyl mercury ingestion.

Regulation and Health

While various studies have shown high concentrations of mercury accumulated in fish, medical cases often go unreported and pose a difficulty in correlating mercury in fish with human poisoning. Environmental issues cover a broad range of areas, but medical cases that are associated with pollutants released into the environment by factories or construction areas cause public health issues that affect not only the environment but also human well-being. Substances poisonous to the human body in a particular amount or dose may not cause any symptoms over time. While there are limits to how much of anything the body can have, mercury is a particular poison that produces immediate physical symptoms when the body has been accumulating it over a period of time.

In the United States, the Environmental Protection Agency measures the amount of mercury concentrated in human blood that does not pose fatal health outcomes. The agency is in charge of enforcing regulations and policies that cover a range of environmental topics. Analysis of blood mercury concentrations in childbearing women has proved that exposure to methyl mercury (MeHg) occurs primarily through the consumption of fish. The U.S. FDA highly recommends against pregnant woman and young children consuming raw fish. Pregnant women and young children often lack strong immune systems and are more at risk for foodborne illnesses.

Medical Cases and Exposure to Mercury

In the United States, the EPA serves as an advisory organ to set the levels of mercury that are non-fatal in humans. Symptoms of exposure to high levels of methyl mercury include disturbed vision, hearing, and speech, lack of coordination, and muscle weakness. Medical studies have examined the correlation of fish consumption and health issues. American studies have presented evidence of fish consumption and its effects on child development. Longitudinal studies agree that human activities are what release and accumulate mercury in marine life. Researchers in the United Kingdom followed a group of children whose mothers ate about 340 or more grams of fish per day. The study concluded that, from the day they were born until they were 42 months old, "children whose mothers ate 340 grams or more had a higher risk and tendency to have lower IQ levels, develop a slowness in motor skills and have a more difficult time developing social skills; As opposed to mothers who did not eat fish or ate very little of it". Addressing the issues of fish consumption forces health officials to recognize the sources of mercury in the human body. Specific Native American tribes are vulnerable to a high exposure of mercury. Studies have determined that these native peoples in the United States suffer more from mercury poisoning and illness than any other cohort group in the country. This is due to the fact that fish is a main source of protein. Exposure risk was assessed through a medical study, thus raising judicial issues of whether the public health of these groups is a priority in the United States.

Work and Exposure

Most cases that arise are due to work exposure or medicinal poisoning. Environmental justice advocates can relate these mercury cases to the unregulated amount of mercury that enters the environment. Workers can be exposed to mercury through the manufacture of fluorescent tubes, chloralkali, or acetaldehyde among other products. Anthropogenic sources and places where mercury is released or used as a solid or vapor puts these has caused fatigue, dizziness, hyperhidrosis, chest congestion, and loss of motor skills. When taken to the hospital, the neurotoxicity levels had already exceeded the maximum amounts. Over-the-counter medicines have been shown to have traces of mercurous choloride. Medical research reported that the children who received doses of these medicines experienced physical symptoms such as "drooling, irregular arm movements, and impaired gait". Exposures to this result in severe physical impairments unregulated chemicals that are put in products. The intake of laxatives that contained about 120 mg of mercurous chloride has also been cases of mercury's toxicity. Two women who abused it over a long period of time took the pills; both died of inorganic mercury poisoning. Both of these raise questions about regulation over products and medicines that go unregulated. Unknown sources of mercury appear to be in the uncommon places, or least expected. Common mercury poisoning cases come from the consumption of fish.

Even in countries, such as Sweden, that have phased out mercury in the dental industry and manufacturing, lingering quantities of mercury still exist in lakes and coastal areas. Moreover, global contributions of mercury to the environment also affect that country. A study in Sweden selected 127 women who had a high level of fish consumption. Around 20% of the women selected, after hair and blood samples, were found to have exceeded the EPA's recommended reference dose of 0.1 micrograms of methyl mercury per kilogram of body weight. Additionally, the study concluded that there was "no margin of safety for neuraldevelopmental effects in fetus[es]" without removing the offending species of fish from the diets of the women. This indicates that families intending to raise children should be especially careful about exposing their unborn babies to toxic mercury via fish.

Children exposed to mercury are particularly susceptible to poisoning since the ratio of food, water, and air intake versus individual body weight is much higher than that of adults. Additionally, children undergo fast growth which causes them to be more susceptible to damaging exposure to methylmercury, as well as the long term consequences of such exposure during childhood development. Young age plays an important role in terms of damage caused by mercury, and much literature on mercury focuses on pregnant women and specific precautions designed to prevent youth mercury exposure. Prenatal methylmercury exposure does cause behavioral problems in infants and worsened cognitive test performance. Additionally, Hughner estimates that 250,000 women may be exposing their unborn babies to levels of methyl mercury above recommended federal levels.

Economically, there does not seem to be a difference in mercury exposure based on socioeconomic bracket and the ability to buy fish from the market. One study shows "no significant differences in mercury levels in tuna, bluefish, and flounder as a function of type of store or economic neighborhood".

By Nation

Certain countries have cultural differences that lead to more fish consumption and therefore more possible exposure to seafood methylmercury. In Ghana, the local population traditionally consumes large quantities of fish, leading to potentially dangerous amounts of mercury in the bloodstream. In the Amazonian Basin, during the rainy season, herbivorous fish dominate the diet of 72.2% of the women selected from a particular Amazonian village. Analysis also shows increase of mercury content in the hair of humans who eat fish on a daily basis in the Amazon.

The most serious case of mercury poisoning in recent history was in the Japanese city of Minamata, in the 1950s. Minamata poisoning proves that significant prenatal and postnatal exposure to high levels of methylmercury causes serious neurological problems. Minamata victims also show higher than normal signs of psychiatric diseases, along with those diseases being caused by underlying neurological issues.

A 2014 USGS survey of mercury levels in the United States water system found that methylmercury concentrations in fish were typically highest in wetland areas including the coastal plain streams in the Southeast. Fish methylmercury levels were also high in the Western US, but only in streams that had been mined for mercury or gold.

Seafood Consumption Benefits

The American College of Obstetricians and Gynecologists note that, considering all the dangers and benefits, the overall result of eating fish in the United States is likely to improve personal health rather than damage it. The college argues that the omega-3 polyunsaturated fatty acids found in fish have a health benefit that outweighs the harm from mercury or polychlorinated biphenyls. Even so, the College also suggests limiting fish consumption for pregnant women. A risk-benefit study weighing the risks of mercury consumption against the benefits derived from fish in Alaska showed that the benefits outweigh the risks when consuming salmon for both cardiovascular health and infant neurological development, and that MeHg data for non-oily fish needs to be of high quality before relative risk can be reliably identified. The Seychelles Child Development Study traced more than seven hundred mother-child pairs for nine years, and found no neurological problems in the children resulting from both prenatal and postnatal methylmercury exposure. A study done with marketed fish in Oman concluded that, except in a few rare cases, the fish available for consumption had lower levels of mercury than limits defined by various health organizations. Clearly, these studies call into question whether or not normal everyday consumption of fish is dangerous in any way, and at very least justify the creation of place-based and culturally relevant consumption advisories. They do not take into account cases of severe mercury poisoning, such as that found in Minamata disease.

Selenium is an element that is known to counteract some of the dangers of ingesting mercury. Multiple studies have been done, such as those in New Jersey and Sweden, that take into account selenium as well as mercury levels. Fish often do contain selenium in conjunction with bioaccumulated mercury, which may offset some of the dangers associated with the mercury ingested.

Current Advice

The complexities associated with mercury transport and environmental fate are described by USEPA in their 1997 Mercury Study Report to Congress. Because methyl mercury and high levels of elemental mercury can be particularly toxic to a fetus or young children, organizations such as the U.S. EPA and FDA recommend that women who are pregnant or plan to become pregnant within the next one or two years, as well as young children, avoid eating more than 6 ounces (170g, one average meal) of fish per week.

In the United States, the FDA has an action level for methylmercury in commercial marine and freshwater fish that is 1.0 parts per million (ppm). In Canada, the limit for the total of mercury content is 0.5 ppm. The Got Mercury? website includes a calculator for determining mercury levels in fish.

Species with characteristically low levels of mercury include shrimp, tilapia, salmon, pollock, and catfish (FDA March 2004). The FDA characterizes shrimp, catfish, pollock, salmon, sardines, and canned light tuna as low-mercury seafood, although recent tests have indicated that up to 6 percent of canned light tuna may contain high levels. A study published in 2008 found that mercury distribution in tuna meat is inversely related to the lipid content, suggesting that the lipid concentration within edible tuna tissues has a diluting effect on mercury content. These findings suggest that choosing to consume a type of tuna that has a higher natural fat content may help reduce the amount of mercury intake, compared to consuming tuna with a low fat content. Also, many of the fish chosen for sushi contain high levels of mercury.

According to the US Food and Drug Administration (FDA), the risk from mercury by eating fish and shellfish is not a health concern for most people. However, certain seafood might contain levels of mercury that may cause harm to an unborn baby (and especially its brain development and nervous system). In a young child, high levels of mercury can interfere with the development of the nervous system. The FDA provides three recommendations for young children, pregnant women, and women of child-bearing age:

1. Do not eat shark, swordfish, king mackerel, or tilefish (Gulf of Mexico) because they might contain high levels of mercury.

2. Eat up to 12 ounces (2 average meals) a week of a variety of fish and shellfish that are lower in mercury. Five of the most commonly eaten fish and shellfish that are low in mercury are: shrimp, canned light tuna, salmon, pollock, and catfish. Another commonly eaten fish, albacore or big eye ("white") tuna depending on its origin might have more mercury than canned light tuna. So, when choosing your two meals of fish and shellfish, it is recommended that you should not eat more than up to 6 ounces (one average meal) of albacore tuna per week.

3. Check local advisories about the safety of fish caught by family and friends in your local lakes, rivers, and coastal areas. If no advice is available, eat up to 6 ounces (one average meal) per week of fish you catch from local waters, but consume no other fish during that week.

Research suggests that selenium content in fish is protective against the toxic effects of methylmercury content. Fish with higher ratios of selenium to methylmercury (Se:Hg) are better to eat since the selenium binds to the methylmercury allowing it to pass through the body un-absorbed.

In 2012 the European Food Safety Authority (EFSA) reported on chemical contaminants they found in the food of over 20 European countries. They established that fish meat and fish products were primarily responsible for methylmercury in the diet of all age classes. Particularly implicated were swordfish, tuna, cod, pike, whiting and hake. The EFSA recommend a tolerable weekly intake for methylmercury of 1.3 µg/kg body weight.

Background

In the 1950s, inhabitants of the seaside town of Minamata, on Kyushu island in Japan, noticed strange behavior in animals. Cats would exhibit nervous tremors, and dance and scream. Within a few years this was observed in other animals; birds would drop out of the sky. Symptoms were also observed in fish, an important component of the diet, especially for the poor. When human symptoms started to be noticed around 1956 an investigation began. Fishing was officially banned in 1957. It was found that the Chisso Corporation, a petrochemical company and maker of plastics such as vinyl chloride, had been discharging heavy metal waste into the sea for decades. They used mercury compounds as catalysts in their syntheses. It is believed that about 5,000 people were killed and perhaps 50,000 have been to some extent poisoned by mercury. Mercury poisoning in Minamata, Japan, is now known as Minamata disease.

Lead Poisoning

Lead poisoning is a type of metal poisoning caused by increased levels of the heavy metal lead in the body. Like most toxic heavy metals, lead interferes with a variety of body processes and is toxic to many organs and tissues, including the heart, bones, intestines, kidneys, and reproductive and nervous systems. The brain is the organ most sensitive to lead exposure. Lead interferes with the development of the nervous system and is therefore particularly toxic to children, causing potentially permanent learning and behavior disorders including violence. Symptoms include abdominal pain, confusion, headache, anemia, irritability, and in severe cases seizures, coma, and death.

Routes of exposure to lead include contaminated air, water, soil, food, and consumer products. Occupational exposure is a common cause of lead poisoning in adults. According to estimates made by the National Institute of Occupational Safety and Health (NIOSH), more than 3 million workers in the United States are potentially exposed to lead in the workplace. One of the largest threats to children is lead paint that exists in many homes, especially older ones; thus children in older housing with chipping paint or lead dust from moveable window frames with lead paint are at greater risk. Prevention of lead exposure can range from individual efforts (e.g., removing lead-containing items such as piping or blinds from the home) to nationwide policies (e.g., laws that ban lead in products, reduce allowable levels in water or soil, or provide for cleanup and mitigation of contaminated soil, etc.)

Elevated lead in the body can be detected by the presence of changes in blood cells visible with a microscope and dense lines in the bones of children seen on X-ray, but the main tool for diagnosis is measurement of the blood lead level. When blood lead levels are recorded, the results indicate how much lead is circulating within the blood stream, not the amount stored in the body. There are two units for reporting blood lead level, either micrograms per deciliter (μg/dl), or micrograms per 100 grams (μg/100 g) of whole blood, which are numerically equivalent. The Centers for Disease Control (US) has set the standard elevated blood lead level for adults to be 10 μg/dl of the whole blood. For children the number is set much lower at 5 μg/dl of blood as of 2012 down from a previous 10 μg/dl. Children are especially prone to the health effects of lead. As a result, blood lead levels must be set lower and closely monitored if contamination is possible. The major treatments are removal of the source of lead and chelation therapy (administration of agents that bind lead so it can be excreted).

Humans have been mining and using this heavy metal for thousands of years, poisoning themselves in the process. Although lead poisoning is one of the oldest known work and environmental hazards, the modern understanding of the small amount of lead necessary to cause harm did not come about until the latter half of the 20th century. No safe threshold for lead exposure has been discovered—that is, there is no known sufficiently small amount of lead that will not cause harm to the body.

Classification

Classically, "lead poisoning" or "lead intoxication" has been defined as exposure to high levels of lead typically associated with severe health effects. Poisoning is a pattern of symptoms that occur with toxic effects from mid to high levels of exposure; toxicity is a wider spectrum of effects, including subclinical ones (those that do not cause symptoms). However, professionals often use "lead poisoning" and "lead toxicity" interchangeably, and official sources do not always restrict the use of "lead poisoning" to refer only to symptomatic effects of lead.

The amount of lead in the blood and tissues, as well as the time course of exposure, determine toxicity. Lead poisoning may be acute (from intense exposure of short duration) or chronic (from repeat low-level exposure over a prolonged period), but the latter is much more common. Diagnosis and treatment of lead exposure are based on blood lead level (the amount of lead in the blood), measured in micrograms of lead per deciliter of blood (µg/dL). Urine lead levels may be used as well, though less commonly. In cases of chronic exposure lead often sequesters in the highest concentrations first in the bones, then in the kidneys. If a provider is performing a provocative excretion test, or "chelation challenge", a measurement obtained from urine rather than blood is likely to provide a more accurate representation of total lead burden to a skilled interpreter.

The US Centers for Disease Control and Prevention and the World Health Organization state that a blood lead level of 10 µg/dL or above is a cause for concern; however, lead may impair development and have harmful health effects even at lower levels, and there is no known safe exposure level. Authorities such as the American Academy of Pediatrics define lead poisoning as blood lead levels higher than 10 µg/dL.

Lead forms a variety of compounds and exists in the environment in various forms. Features of poisoning differ depending on whether the agent is an organic compound (one that contains carbon), or an inorganic one. Organic lead poisoning is now very rare, because countries across the world have phased out the use of organic lead compounds as gasoline additives, but such compounds are still used in industrial settings. Organic lead compounds, which cross the skin and respiratory tract easily, affect the central nervous system predominantly.

Signs and Symptoms

Lead poisoning can cause a variety of symptoms and signs which vary depending on the individual and the duration of lead exposure. Symptoms are nonspecific and may be subtle, and someone with elevated lead levels may have no symptoms. Symptoms usually develop over weeks to months as lead builds up in the body during a chronic exposure, but acute symptoms from brief, intense exposures also occur. Symptoms from exposure to organic lead, which is probably more toxic than inorganic lead due to its lipid solubility, occur rapidly. Poisoning by organic lead compounds has symptoms predominantly in the central nervous system, such as insomnia, delirium, cognitive deficits, tremor, hallucinations, and convulsions.

Symptoms may be different in adults and children; the main symptoms in adults are headache, abdominal pain, memory loss, kidney failure, male reproductive problems, and weakness, pain, or tingling in the extremities.

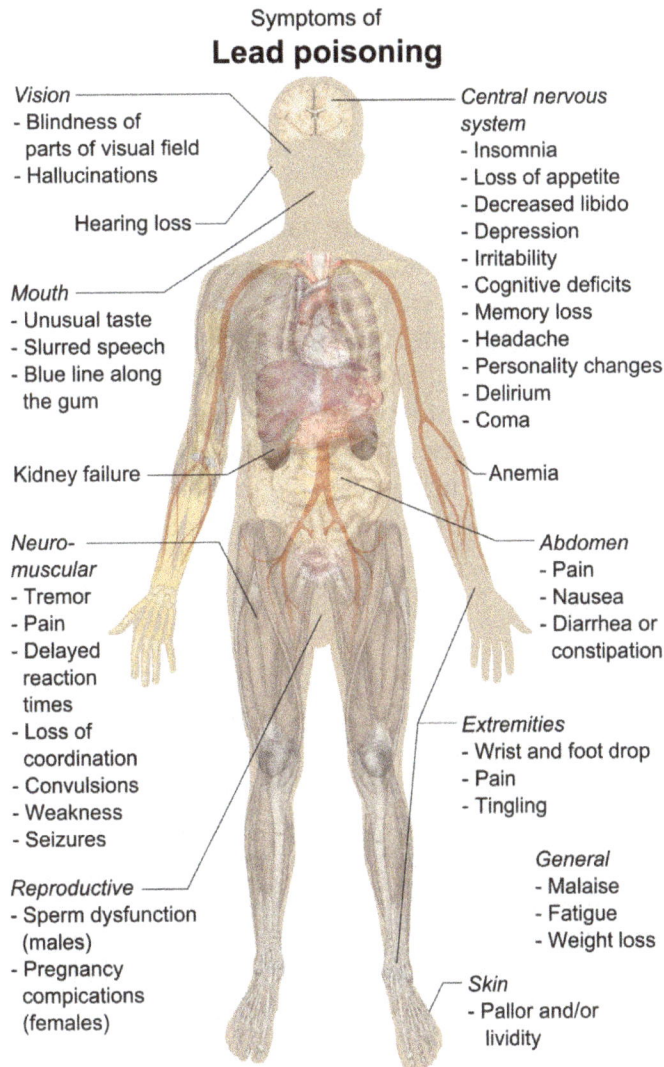

Symptoms of
Lead poisoning

Vision
- Blindness of
  parts of visual field
- Hallucinations

Hearing loss

Mouth
- Unusual taste
- Slurred speech
- Blue line along
  the gum

Kidney failure

*Neuro-
muscular*
- Tremor
- Pain
- Delayed
  reaction
  times
- Loss of
  coordination
- Convulsions
- Weakness
- Seizures

Reproductive
- Sperm dysfunction
  (males)
- Pregnancy
  compications
  (females)

*Central nervous
system*
- Insomnia
- Loss of appetite
- Decreased libido
- Depression
- Irritability
- Cognitive deficits
- Memory loss
- Headache
- Personality changes
- Delirium
- Coma

Anemia

Abdomen
- Pain
- Nausea
- Diarrhea or
  constipation

Extremities
- Wrist and foot drop
- Pain
- Tingling

General
- Malaise
- Fatigue
- Weight loss

Skin
- Pallor and/or
  lividity

Symptoms of lead poisoning.

Early symptoms of lead poisoning in adults are commonly nonspecific and include depression, loss of appetite, intermittent abdominal pain, nausea, diarrhea, constipation, and muscle pain. Other early signs in adults include malaise, fatigue, decreased libido, and problems with sleep. An unusual taste in the mouth and personality changes are also early signs.

In adults, symptoms can occur at levels above 40 µg/dL, but are more likely to occur only above 50–60 µg/dL. Symptoms begin to appear in children generally at around 60 µg/dL. However, the lead levels at which symptoms appear vary widely depending on unknown characteristics of each individual. At blood lead levels between 25 and 60 µg/dL, neuropsychiatric effects such as delayed reaction times, irritability, and difficulty concentrating, as well as slowed motor nerve conduction and headache can occur. Anemia may appear at blood lead levels higher than 50 µg/dL. In adults, abdominal colic, involving paroxysms of pain, may appear at blood lead levels greater than 80 µg/dL. Signs that occur in adults at blood lead levels exceeding 100 µg/dL include wrist drop and foot drop, and signs of encephalopathy (a condition characterized by brain swelling), such as those that accompany increased pressure within the skull, delirium, coma, seizures, and headache. In children, signs of encephalopathy such as bizarre behavior, discoordination, and apathy occur at lead levels exceeding 70 µg/dL. For both adults and children, it is rare to be asymptomatic if blood lead levels exceed 100 µg/dL.

Acute Poisoning

In acute poisoning, typical neurological signs are pain, muscle weakness, numbness and tingling, and, rarely, symptoms associated with inflammation of the brain. Abdominal pain, nausea, vomiting, diarrhea, and constipation are other acute symptoms. Lead's effects on the mouth include astringency and a metallic taste. Gastrointestinal problems, such as constipation, diarrhea, poor appetite, or weight loss, are common in acute poisoning. Absorption of large amounts of lead over a short time can cause shock (insufficient fluid in the circulatory system) due to loss of water from the gastrointestinal tract. Hemolysis (the rupture of red blood cells) due to acute poisoning can cause anemia and hemoglobin in the urine. Damage to kidneys can cause changes in urination such as decreased urine output. People who survive acute poisoning often go on to display symptoms of chronic poisoning.

Chronic Poisoning

Chronic poisoning usually presents with symptoms affecting multiple systems, but is associated with three main types of symptoms: gastrointestinal, neuromuscular, and neurological. Central nervous system and neuromuscular symptoms usually result from intense exposure, while gastrointestinal symptoms usually result from exposure over longer periods. Signs of chronic exposure include loss of short-term memory or concentration, depression, nausea, abdominal pain, loss of coordination, and numbness and tingling in the extremities. Fatigue, problems with sleep, headaches, stupor, slurred speech, and anemia are also found in chronic lead poisoning. A "lead hue" of the skin with pallor and/or lividity is another feature. A blue line along the gum with bluish black edging to the teeth, known as a Burton line, is another indication of chronic lead poisoning. Children with chronic poisoning may refuse to play or may have hyperkinetic or aggressive behavior disorders. Visual disturbance may present with gradually progressing blurred vision as a result of central scotoma, caused by toxic optic neuritis.

Effects on Children

A fetus developing in the womb of a woman who has elevated blood lead level is susceptible to lead poisoning by intrauterine exposure, and is at greater risk of being born prematurely or with a low birth weight.

Children are more at risk for lead poisoning because their smaller bodies are in a continuous state of growth and development. Lead is absorbed at a faster rate compared to adults, which causes more physical harm than to older people. Furthermore, children, especially as they are learning to crawl and walk, are constantly on the floor and therefore more prone to ingesting and inhaling dust that is contaminated with lead.

The classic signs and symptoms in children are loss of appetite, abdominal pain, vomiting, weight loss, constipation, anemia, kidney failure, irritability, lethargy, learning disabilities, and behavioral problems. Slow development of normal childhood behaviors, such as talking and use of words, and permanent intellectual disability are both commonly seen. Although less common, it is possible for fingernails to develop leukonychia striata if exposed to abnormally high lead concentrations.

Complications

Lead affects every one of the body's organ systems, especially the nervous system, but also the bones and teeth, the kidneys, and the cardiovascular, immune, and reproductive systems. Hearing loss and tooth decay have been linked to lead exposure, as have cataracts. Intrauterine and neo-natal lead exposure promote tooth decay. Aside from the developmental effects unique to young children, the health effects experienced by adults are similar to those in children, although the thresholds are generally higher.

Kidneys

Kidney damage occurs with exposure to high levels of lead, and evidence suggests that lower levels can damage kidneys as well. The toxic effect of lead causes nephropathy and may cause Fanconi syndrome, in which the proximal tubular function of the kidney is impaired. Long-term exposure at levels lower than those that cause lead nephropathy have also been reported as nephrotoxic in patients from developed countries that had chronic kidney disease or were at risk because of hypertension or diabetes mellitus. Lead poisoning inhibits excretion of the waste product urate and causes a predisposition for gout, in which urate builds up. This condition is known as *saturnine gout*.

Cardiovascular System

Evidence suggests lead exposure is associated with high blood pressure, and studies have also found connections between lead exposure and coronary heart disease, heart rate variability, and death from stroke, but this evidence is more limited. People who have been exposed to higher concentrations of lead may be at a higher risk for cardiac autonomic dysfunction on days when ozone and fine particles are higher.

Reproductive System

Lead affects both the male and female reproductive systems. In men, when blood lead levels exceed 40 µg/dL, sperm count is reduced and changes occur in volume of sperm, their motility, and their morphology. A pregnant woman's elevated blood lead level can lead to miscarriage, prematurity, low birth weight, and problems with development during childhood. Lead is able to pass through the placenta and into breast milk, and blood lead levels in mothers and infants are usually similar. A fetus may be poisoned *in utero* if lead from the mother's bones is subsequently mobilized by the changes in metabolism due to pregnancy; increased calcium intake in pregnancy may help mitigate this phenomenon.

Nervous System

Lead affects the peripheral nervous system (especially motor nerves) and the central nervous system. Peripheral nervous system effects are more prominent in adults and central nervous system effects are more prominent in children. Lead causes the axons of nerve cells to degenerate and lose their myelin coats.

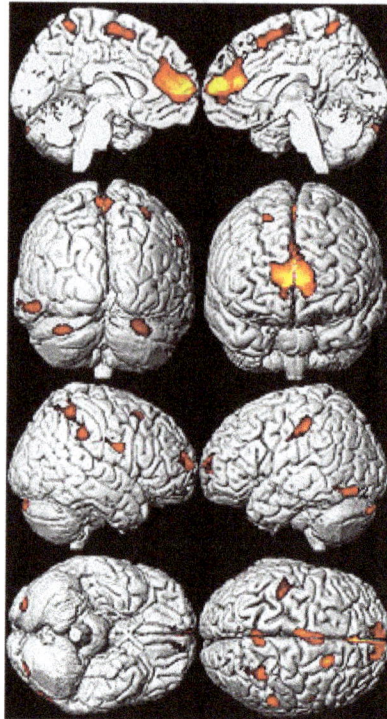

The brains of adults who were exposed to lead as children show decreased volume, especially in the prefrontal cortex, on MRI. Areas of volume loss are shown in color over a template of a normal brain.

Lead exposure in young children has been linked to learning disabilities, and children with blood lead concentrations greater than 10 µg/dL are in danger of developmental disabilities. Increased blood lead level in children has been correlated with decreases in intelligence, nonverbal reasoning, short-term memory, attention, reading and arithmetic ability, fine motor skills, emotional regulation, and social engagement. The effect of lead on children's cognitive abilities takes place at very low levels. There is apparently no lower threshold to the dose-response relationship (unlike other heavy metals such as mercury). Reduced academic performance has been associated with lead exposure even at blood lead levels lower than 5 µg/dL. Blood lead levels below 10 µg/dL have been reported to be associated with lower IQ and behavior problems such as aggression, in proportion with blood lead levels. Between the blood lead levels of 5 and 35 µg/dL, an IQ decrease of 2–4 points for each µg/dL increase is reported in children.

High blood lead levels in adults are also associated with decreases in cognitive performance and with psychiatric symptoms such as depression and anxiety. It was found in a large group of current and former inorganic lead workers in Korea that blood lead levels in the range of 20–50 µg/dL were correlated with neuro-cognitive defects. Increases in blood lead levels from about 50 to about 100 µg/dL in adults have been found to be associated with persistent, and possibly permanent, impairment of central nervous system function.

Lead exposure in children is also correlated with neuropsychiatric disorders such as attention deficit hyperactivity disorder and anti-social behaviour. Elevated lead levels in children are correlated with higher scores on aggression and delinquency measures. A correlation has also been found between prenatal and early childhood lead exposure and violent crime in adulthood. Countries with the highest air lead levels have also been found to have the highest murder rates, after adjusting for confounding factors. A May 2000 study by economic consultant Rick Nevin theorizes that lead exposure explains 65% to 90% of the variation in violent crime rates in the US. A 2007 paper by the same author claims to show a strong association between preschool blood lead and subsequent crime rate trends over several decades across nine countries. It is believed that the U.S. ban on lead paint in buildings in the late 1970s, as well as the phaseout of leaded gasoline in the 1970s and 1980s, partially helped contribute to the decline of violent crime in the United States since the early 1990s.

Exposure Routes

Lead is a common environmental pollutant. Causes of environmental contamination include industrial use of lead, such as is found in facilities that process lead-acid batteries or produce lead wire or pipes, and metal recycling and foundries. Children living near facilities that process lead, such as lead smelters, have been found to have unusually high blood lead levels. In August 2009, parents rioted in China after lead poisoning was found in nearly 2000 children living near zinc and manganese smelters. Lead exposure can occur from contact with lead in air, household dust, soil, water, and commercial products. Leaded gasoline has also been linked to increases in lead pollution. Some research has suggested a link between leaded gasoline and crime rates.

Occupational Exposure

Battery recycling workers are at risk for lead exposure. This worker ladles molten lead into billets in a lead-acid battery recovery facility.

In adults, occupational exposure is the main cause of lead poisoning. People can be exposed when working in facilities that produce a variety of lead-containing products; these include radiation shields, ammunition, certain surgical equipment, developing dental x-ray films prior to digital x-rays (each film packet had a lead liner to prevent the radiation from going through), fetal monitors, plumbing, circuit boards, jet engines, and ceramic glazes. In addition, lead miners and smelters, plumbers and fitters, auto mechanics, glass manufacturers, construction workers, battery manufacturers and recyclers, firing range instructors, and plastic manufacturers are at risk for lead exposure. Other occupations that present lead exposure risks include welding, manufacture of rubber, printing, zinc and copper smelting, processing of ore, combustion of solid waste, and production of paints and pigments. Parents who are exposed to lead in the workplace can bring lead dust home on clothes or skin and expose their children.

Food

Lead may be found in food when food is grown in soil that is high in lead, airborne lead contaminates the crops, animals eat lead in their diet, or lead enters the food either from what it was stored or cooked in.

Paint

Some lead compounds are colorful and are used widely in paints, and lead paint is a major route of lead exposure in children. A study conducted in 1998–2000 found that 38 million housing units in the US had lead-based paint, down from a 1990 estimate of 64 million. Deteriorating lead paint can produce dangerous lead levels in household dust and soil. Deteriorating lead paint and lead-containing household dust are the main causes of chronic lead poisoning. The lead breaks down into the dust and since children are more prone to crawling on the floor, it is easily ingested. Many young children display pica, eating things that are not food. Even a small amount of a lead-containing product such as a paint chip or a sip of glaze can contain tens or hundreds of milligrams of lead. Eating chips of lead paint presents a particular hazard to children, generally producing more severe poisoning than occurs from dust. Because removing lead paint from dwellings, e.g. by sanding or torching creates lead-containing dust and fumes, it is generally safer to seal the lead paint under new paint (excepting moveable windows and doors, which create paint dust when operated). Alternately, special precautions must be taken if the lead paint is to be removed. In oil painting it was once common for colours such as yellow or white to be made with lead carbonate. Lead white oil colour was the main white of oil painters until superseded by compounds containing zinc or titanium in the mid-20th century. It is speculated that the painter Caravaggio and possibly Francisco Goya and Vincent Van Gogh had lead poisoning due to overexposure or carelessness when handling this colour.

Soil

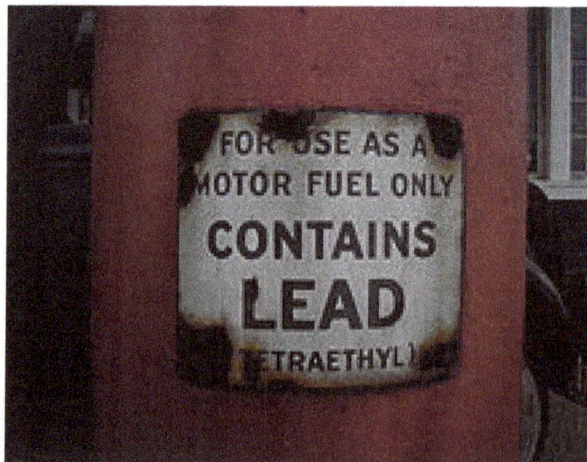

A lead warning on a fuel pump. Tetraethyllead, which used to be added to automotive gasoline (and still is added to some aviation gasolines), contributed to soil contamination.

Residual lead in soil contributes to lead exposure in urban areas. It has been thought that the more polluted an area is with various contaminants, the more likely it is to contain lead. However, this is not always the case, as there are several other reasons for lead contamination in soil. Lead content in soil may be caused by broken-down lead paint, residues from lead-containing gasoline, used engine oil, or pesticides used in the past, contaminated landfills, or from nearby industries such as foundries or smelters. Although leaded soil is less of a problem in countries that no longer have leaded gasoline, it remains prevalent, raising concerns about the safety of urban agriculture; eating food grown in contaminated soil can present a lead hazard.

Water

Lead from the atmosphere or soil can end up in groundwater and surface water. It is also potentially in drinking water, e.g. from plumbing and fixtures that are either made of lead or have lead solder. Since acidic water breaks down lead in plumbing more readily, chemicals can be added to municipal water to increase the pH and thus reduce the corrosivity of the public water supply. Chloramines, which were adopted as a substitute for chlorine disinfectants due to fewer health concerns, increase corrositivity. In the US, 14–20% of total lead exposure is attributed to drinking water. In 2004, a team of seven reporters from *The Washington Post* discovered high levels of lead in the drinking water in Washington, D.C. and won an award for investigative reporting for a series of articles about this contamination. In the Flint water crisis, a switch to a more corrosive municipal water source elevated lead levels in drinking water in domestic tap water.

In Australia, collecting rainwater from roof runoff used as potable water may contain lead if there are lead contaminants on the roof or in the storage tank. The Australian Drinking Water Guidelines allow a maximum of .01 mg/L lead in water.

Lead-containing Products

Lead can be found in products such as kohl, an ancient cosmetic from the Middle East, South Asia, and parts of Africa that has many names; and from some toys. In 2007, millions of toys made in China were recalled from multiple countries owing to safety hazards including lead paint. Vinyl mini-blinds, found especially in older housing, may contain lead. Lead is commonly incorporated into herbal remedies such as Indian Ayurvedic preparations and remedies of Chinese origin. There are also risks of elevated blood lead levels caused by folk remedies like *azarcon* and *greta*, which each contain about 95% lead.

Ingestion of metallic lead, such as small lead fishing lures, increases blood lead levels and can be fatal. Ingestion of lead-contaminated food is also a threat. Ceramic glaze often contains lead, and dishes that have been improperly fired can leach the metal into food, potentially causing severe poisoning. In some places, the solder in cans used for food contains lead. When manufacturing medical instruments and hardware, solder containing lead may be present. People who eat animals hunted with lead bullets may be at risk for lead exposure. Bullets lodged in the body rarely cause significant levels of lead, but bullets lodged in the joints are the exception, as they deteriorate and release lead into the body over time.

In May 2015, Indian food safety regulators in the state of Uttar Pradesh found that samples of Maggi 2 Minute Noodles contained lead up to 17 times beyond permissible limits. On 3 June 2015,

New Delhi Government banned the sale of Maggi noodles in New Delhi stores for 15 days because it was found to contain lead beyond the permissible limit. The Gujarat FDA on June 4, 2015 banned the noodles for 30 days after 27 out of 39 samples were detected with objectionable levels of metallic lead, among other things. Some India's biggest retailers like Future Group, Big Bazaar, Easyday and Nilgiris have imposed a nationwide ban on Maggi noodles. Many other states too have banned Maggi noodles.

Bullets

Contact with ammunition is a source of lead exposure. As of 2013, lead-based ammunition production is the second largest annual use of lead in the US, accounting for over 60,000 metric tons consumed in 2012, second only to the manufacture of storage batteries. The Environmental Protection Agency (EPA) does not regulate lead, as a matter of law. Lead birdshot is banned in some areas, but this is primarily for the benefit of the birds and their predators, rather than humans. Non-lead alternatives include steel, tungsten-nickel-iron, bismuth-tin, and tungsten-polymer.

Because game animals can be shot using lead bullets, the potential for lead ingestion from game meat consumption has been studied clinically and epidemiologically. In a recent study conducted by the CDC, a cohort from North Dakota was enrolled and asked to self-report historical consumption of game meat, and participation in other activities that could cause lead exposure. The study found that participants' age, sex, housing age, current hobbies with potential for lead exposure, and game consumption were all associated with blood lead level (PbB).

This study has been cited by popular media as simple evidence that hunting increases exposure to lead poisoning, prompting the University of Illinois Extension to release a statement that there is no such risk. Concerning the CDC report, the authors' conclusion in a related Epi-AID Trip Report notes the small increase associated with game consumption in the study, and urges interpretation with respect to environmental context:

While this study suggests that consumption of wild game meat can adversely affect PbB, no participant had PbB higher than the CDC recommended threshold of 10µg/dl—the level at which CDC recommends case management; and the geometric mean PbB among this study population (1.17µg/dl) was lower than the overall population geometric mean PbB in the United States (1.60 µg/dl). The clinical significance of low PbB in this sample population and the small quantitative increase of 0.30µg/dl in PbB associated with wild game consumption should be interpreted in the context of naturally occurring PbB.

Copper-jacketed, lead-based bullets are more economical to produce and use than lead or any other material. Alternative materials are available such as steel, copper, and tungsten, but alternatives are universally less effective and/or more expensive. However, the biggest impediment to using the vast majority of alternatives relates to current laws in the United States pertaining to armor-piercing rounds. Laws and regulations relating to armor-piercing ammunition expressly prohibit the use of brass, bronze, steel, tungsten, and nearly every metallic alternative in any bullet that can be shot by a handgun, which at this time is nearly every caliber smaller than 50BMG (including the popular .223 Remington, .308 Winchester and .30-06 to name just a few). Some lead-based bullets are resistant to fragmentation, offering hunters the ability to clean game animals with negligible risk of including lead fragments in prepared meat. Other bullets are prone to fragmentation and exacerbate the risk of lead ingestion from prepared meat. In practice, use of a non-fragmenting bullet and proper cleaning of the game animal's wound can eliminate the risk of lead ingestion from eating game; however, isolating such practice to experimentally determine its association with blood lead levels in study is difficult. Bismuth is an element used as a lead-replacement for shotgun pellets used in waterfowl hunting although shotshells made from bismuth are nearly ten times the cost of lead.

Pathophysiology

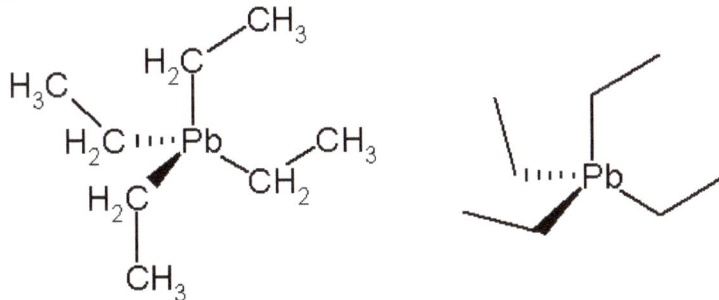

Tetraethyllead, still used as an additive in some fuels, can be absorbed through the skin.

Exposure occurs through inhalation, ingestion or occasionally skin contact. Lead may be taken in through direct contact with mouth, nose, and eyes (mucous membranes), and through breaks in the skin. Tetraethyllead, which was a gasoline additive and is still used in fuels such as aviation fuel, passes through the skin; however inorganic lead found in paint, food, and most lead-containing consumer products is only minimally absorbed through the skin. The main sources of absorption of inorganic lead are from ingestion and inhalation. In adults, about 35–40% of inhaled lead dust is deposited in the lungs, and about 95% of that goes into the bloodstream. Of ingested inorganic lead, about 15% is absorbed, but this percentage is higher in children, pregnant women, and people with deficiencies of calcium, zinc, or iron. Children and infants may absorb about 50% of ingested lead, but little is known about absorption rates in children.

The main body compartments that store lead are the blood, soft tissues, and bone; the half-life of lead in these tissues is measured in weeks for blood, months for soft tissues, and years for bone. Lead in the bones, teeth, hair, and nails is bound tightly and not available to other tissues, and is generally thought not to be harmful. In adults, 94% of absorbed lead is deposited in the bones and teeth, but children only store 70% in this manner, a fact which may partially account for the more serious health effects on children. The estimated half-life of lead in bone is 20 to 30 years, and bone can introduce lead into the bloodstream long after the initial exposure is gone. The half-life of lead in the blood in men is about 40 days, but it may be longer in children and pregnant women, whose bones are undergoing remodeling, which allows the lead to be continuously re-introduced into the bloodstream. Also, if lead exposure takes place over years, clearance is much slower, partly due to the re-release of lead from bone. Many other tissues store lead, but those with the highest concentrations (other than blood, bone, and teeth) are the brain, spleen, kidneys, liver, and lungs. It is removed from the body very slowly, mainly through urine. Smaller amounts of lead are also eliminated through the feces, and very small amounts in hair, nails, and sweat.

Lead has no known physiologically relevant role in the body, and its harmful effects are myriad. Lead and other heavy metals create reactive radicals which damage cell structures including DNA and cell membranes. Lead also interferes with DNA transcription, enzymes that help in the synthesis of vitamin D, and enzymes that maintain the integrity of the cell membrane. Anemia may result when the cell membranes of red blood cells become more fragile as the result of damage to their membranes. Lead interferes with metabolism of bones and teeth and alters the permeability of blood vessels and collagen synthesis. Lead may also be harmful to the developing immune system, causing production of excessive inflammatory proteins; this mechanism may mean that lead exposure is a risk factor for asthma in children. Lead exposure has also been associated with a decrease in activity of immune cells such as polymorphonuclear leukocytes. Lead also interferes with the normal metabolism of calcium in cells and causes it to build up within them.

Enzymes

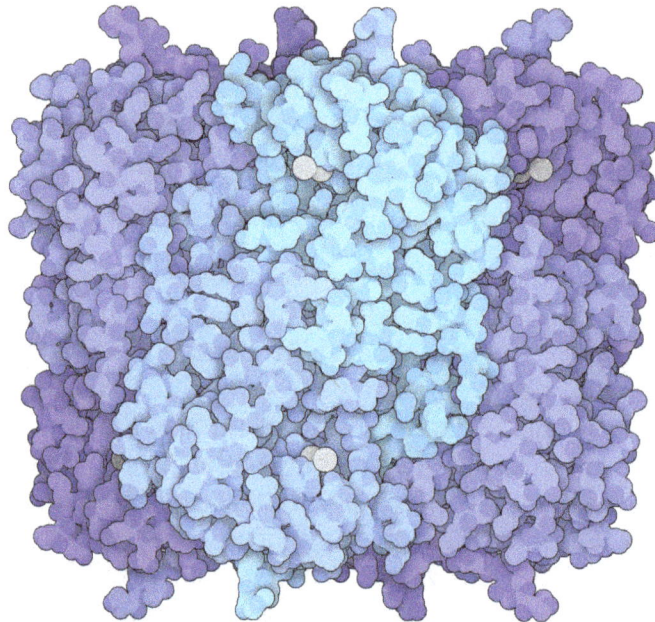

ALAD enzyme with lead bound (PDB: 1QNV)

The primary cause of lead's toxicity is its interference with a variety of enzymes because it binds to sulfhydryl groups found on many enzymes. Part of lead's toxicity results from its ability to mimic other metals that take part in biological processes, which act as cofactors in many enzymatic reactions, displacing them at the enzymes on which they act. Lead is able to bind to and interact with many of the same enzymes as these metals but, due to its differing chemistry, does not properly function as a cofactor, thus interfering with the enzyme's ability to catalyze its normal reaction or reactions. Among the essential metals with which lead interacts are calcium, iron, and zinc.

One of the main causes for the pathology of lead is that it interferes with the activity of an essential enzyme called delta-aminolevulinic acid dehydratase, or ALAD, which is important in the biosynthesis of heme, the cofactor found in hemoglobin. Lead also inhibits the enzyme ferrochelatase, another enzyme involved in the formation of heme. Ferrochelatase catalyzes the joining of protoporphyrin and Fe^{2+} to form heme. Lead's interference with heme synthesis results in production of zinc protoporphyrin and the development of anemia. Another effect of lead's interference with heme synthesis is the buildup of heme precursors, such as aminolevulinic acid, which may be directly or indirectly harmful to neurons.

Neurons

The brain is the organ most sensitive to lead exposure. Lead is able to pass through the endothelial cells at the blood brain barrier because it can substitute for calcium ions and be uptaken by Calcium-ATPase pumps. Lead poisoning interferes with the normal development of a child's brain and nervous system; therefore children are at greater risk of lead neurotoxicity than adults are. In a child's developing brain, lead interferes with synapse formation in the cerebral cortex, neurochemical development (including that of neurotransmitters), and organization of ion channels. It causes loss of neurons' myelin sheaths, reduces numbers of neurons, interferes with neurotransmission, and decreases neuronal growth.

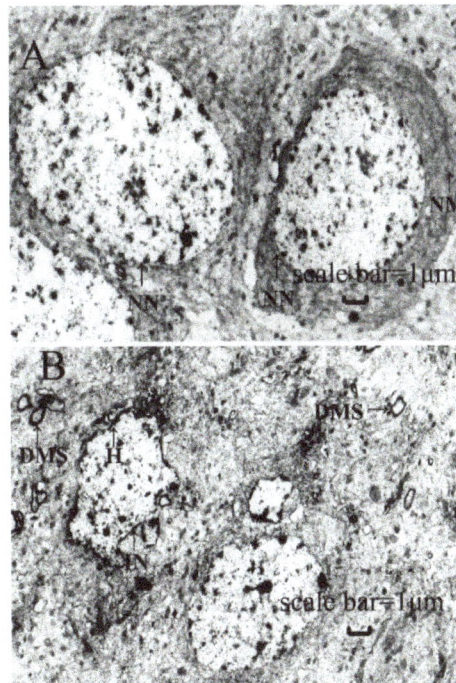

Lead exposure damages cells in the hippocampus, a part of the brain involved in memory. Hippocampi of lead-exposed rats (bottom) show structural damage such as irregular nuclei (IN) and denaturation of myelin (DNS) compared to controls (top).

Lead interferes with the release of neurotransmitters, chemicals used by neurons to send signals to other cells. It interferes with the release of glutamate, a neurotransmitter important in many functions including learning, by blocking NMDA receptors. The targeting of NMDA receptors is thought to be one of the main causes for lead's toxicity to neurons. A Johns Hopkins University report found that in addition to inhibiting the NMDA receptor, lead exposure decreased the level of expression of the gene for the receptor in part of the brain. In addition, lead has been found in animal studies to cause programmed cell death in brain cells.

Diagnosis

Diagnosis includes determining the clinical signs and the medical history, with inquiry into possible routes of exposure. Clinical toxicologists, medical specialists in the area of poisoning, may be involved in diagnosis and treatment. The main tool in diagnosing and assessing the severity of lead poisoning is laboratory analysis of the blood lead level (BLL).

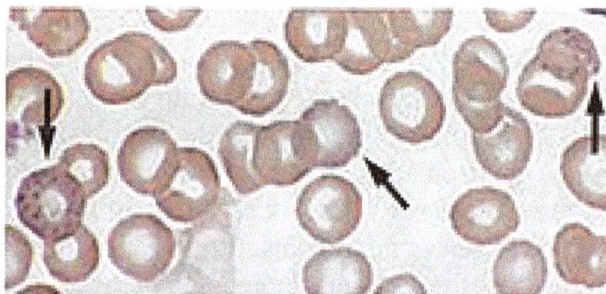

Basophilic stippling (arrows) of red blood cells in a 53-year-old who had elevated blood lead levels due to drinking repeatedly from glasses decorated with lead paint.

Blood film examination may reveal basophilic stippling of red blood cells (dots in red blood cells visible through a microscope), as well as the changes normally associated with iron-deficiency anemia (microcytosis and hypochromasia). However, basophilic stippling is also seen in unrelated conditions, such as megaloblastic anemia caused by vitamin B12 (colbalamin) and folate deficiencies.

Exposure to lead also can be evaluated by measuring erythrocyte protoporphyrin (EP) in blood samples. EP is a part of red blood cells known to increase when the amount of lead in the blood is high, with a delay of a few weeks. Thus EP levels in conjunction with blood lead levels can suggest the time period of exposure; if blood lead levels are high but EP is still normal, this finding suggests exposure was recent. However, the EP level alone is not sensitive enough to identify elevated blood lead levels below about 35 μg/dL. Due to this higher threshold for detection and the fact that EP levels also increase in iron deficiency, use of this method for detecting lead exposure has decreased.

Blood lead levels are an indicator mainly of recent or current lead exposure, not of total body burden. Lead in bones can be measured noninvasively by X-ray fluorescence; this may be the best measure of cumulative exposure and total body burden. However this method is not widely available and is mainly used for research rather than routine diagnosis. Another radiographic sign of elevated lead levels is the presence of radiodense lines called lead lines at the metaphysis in the long bones of growing children, especially around the knees. These lead lines, caused by increased calcification due to disrupted metabolism in the growing bones, become wider as the duration of lead exposure increases. X-rays may also reveal lead-containing foreign materials such as paint chips in the gastrointestinal tract.

Fecal lead content that is measured over the course of a few days may also be an accurate way to estimate the overall amount of childhood lead intake. This form of measurement may serve as a useful way to see the extent of oral lead exposure from all the diet and environmental sources of lead.

Lead poisoning shares symptoms with other conditions and may be easily missed. Conditions that present similarly and must be ruled out in diagnosing lead poisoning include carpal tunnel syndrome, Guillain-Barré syndrome, renal colic, appendicitis, encephalitis in adults, and viral gastroenteritis in children. Other differential diagnoses in children include constipation, abdominal colic, iron deficiency, subdural hematoma, neoplasms of the central nervous system, emotional and behavior disorders, and intellectual disability.

Reference Levels

The current reference range for acceptable blood lead concentrations in healthy persons without excessive exposure to environmental sources of lead is less than 5 µg/dL for children. It was less than 25 µg/dL for adults. Previous to 2012 the value for children was 10 (µg/dl). The current biological exposure index (a level that should not be exceeded) for lead-exposed workers in the U.S. is 30 µg/dL in a random blood specimen.

The National Institute for Occupational Safety and Health (CDC/NIOSH) reference blood lead level in adults is 10 µg/dL The U.S. national BLL geometric mean among adults was 1.2 µg/dL in 2009–2010.

Blood lead concentrations in poisoning victims have ranged from 30->80 µg/dL in children exposed to lead paint in older houses, 77–104 µg/dL in persons working with pottery glazes, 90–137 µg/dL in individuals consuming contaminated herbal medicines, 109–139 µg/dL in indoor shooting range instructors and as high as 330 µg/dL in those drinking fruit juices from glazed earthenware containers.

Prevention

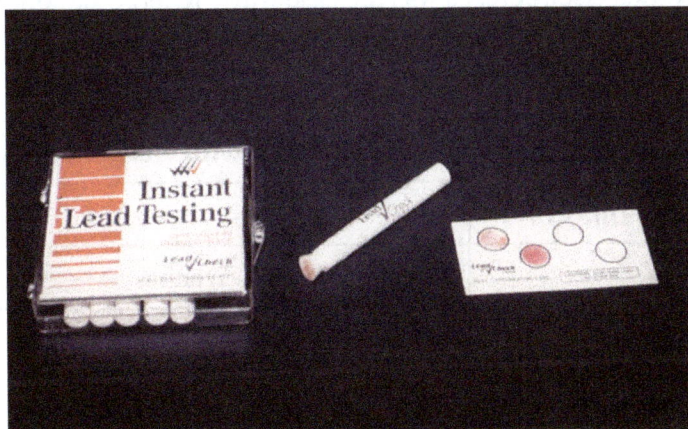

Testing kits are commercially available for detecting lead. These swabs, when wiped on a surface, turn red in the presence of lead.

In most cases, lead poisoning is preventable by avoiding exposure to lead. Prevention strategies can be divided into individual (measures taken by a family), preventive medicine (identifying and intervening with high-risk individuals), and public health (reducing risk on a population level).

Recommended steps by individuals to reduce the blood lead levels of children include increasing their frequency of hand washing and their intake of calcium and iron, discouraging them from putting their hands to their mouths, vacuuming frequently, and eliminating the presence of lead-containing objects such as blinds and jewellery in the house. In houses with lead pipes or plumbing solder, these can be replaced. Less permanent but cheaper methods include running water in the morning to flush out the most contaminated water, or adjusting the water's chemistry to prevent corrosion of pipes. Lead testing kits are commercially available for detecting the presence of lead in the household. As hot water is more likely than cold water to contain higher amounts of lead, use only cold water from the tap for drinking, cooking, and for making baby formula. Since most of the lead in household water usually comes from plumbing in the house and not from the local water supply, using cold water can avoid lead exposure. Measures such as dust control and household education do not appear to be effective in changing children's blood levels.

Screening is an important method in preventive medicine strategies. Screening programs exist to test the blood of children at high risk for lead exposure, such as those who live near lead-related industries.

Prevention measures also exist on national and municipal levels. Recommendations by health professionals for lowering childhood exposures include banning the use of lead where it is not essential and strengthening regulations that limit the amount of lead in soil, water, air, household dust, and products. Regulations exist to limit the amount of lead in paint; for example, a 1978 law in the US restricted the lead in paint for residences, furniture, and toys to 0.06% or less. In October 2008, the US Environmental Protection Agency reduced the allowable lead level by a factor of ten to 0.15 micrograms per cubic meter of air, giving states five years to comply with the standards. The European Union's Restriction of Hazardous Substances Directive limits amounts of lead and other toxic substances in electronics and electrical equipment. In some places, remediation programs exist to reduce the presence of lead when it is found to be high, for example in drinking water. As a more radical solution, entire towns located near former lead mines have been "closed" by the government, and the population resettled elsewhere, as was the case with Picher, Oklahoma in 2009.

Treatment

CDC management guidelines for children with elevated blood levels	
Blood lead level (µg/dL)	**Treatment**
10–14	Education, repeat screening
15–19	Repeat screening, case management to abate sources
20–44	Medical evaluation, case management
45–69	Medical evaluation, chelation, case management
>69	Hospitalization, immediate chelation, case management

The mainstays of treatment are removal from the source of lead and, for people who have significantly high blood lead levels or who have symptoms of poisoning, chelation therapy. Treatment of iron, calcium, and zinc deficiencies, which are associated with increased lead absorption, is another part of treatment for lead poisoning. When lead-containing materials are present in the gastrointestinal tract (as evidenced by abdominal X-rays), whole bowel irrigation, cathartics, endoscopy, or even surgical removal may be used to eliminate it from the gut and prevent further exposure. Lead-containing bullets and shrapnel may also present a threat of further exposure and may need to be surgically removed if they are in or near fluid-filled or synovial spaces. If lead encephalopathy is present, anticonvulsants may be given to control seizures, and treatments to control swelling of the brain include corticosteroids and mannitol. Treatment of organic lead poisoning involves removing the lead compound from the skin, preventing further exposure, treating seizures, and possibly chelation therapy for people with high blood lead concentrations.

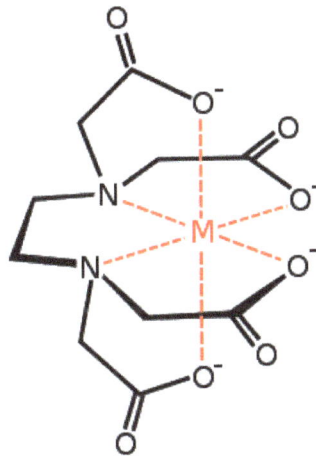

EDTA, a chelating agent, binds a heavy metal, sequestering it.

A chelating agent is a molecule with at least two negatively charged groups that allow it to form complexes with metal ions with multiple positive charges, such as lead. The chelate that is thus formed is nontoxic and can be excreted in the urine, initially at up to 50 times the normal rate. The chelating agents used for treatment of lead poisoning are edetate disodium calcium (CaNa$_2$EDTA), dimercaprol (BAL), which are injected, and succimer and d-penicillamine, which are administered orally. Chelation therapy is used in cases of acute lead poisoning, severe poisoning, and encephalopathy, and is considered for people with blood lead levels above 25 µg/dL. While the use of chelation for people with symptoms of lead poisoning is widely supported, use in asymptomatic people with high blood lead levels is more controversial. Chelation therapy is of limited value for cases of chronic exposure to low levels of lead. Chelation therapy is usually stopped when symptoms resolve or when blood lead levels return to premorbid levels. When lead exposure has taken place over a long period, blood lead levels may rise after chelation is stopped because lead is leached into blood from stores in the bone; thus repeated treatments are often necessary.

People receiving dimercaprol need to be assessed for peanut allergies since the commercial formulation contains peanut oil. Calcium EDTA is also effective if administered four hours after the administration of dimercaprol. Administering dimercaprol, DMSA (Succimer), or DMPS prior to calcium EDTA is necessary to prevent the redistribution of lead into the central nervous system. Dimercaprol used alone may also redistribute lead to the brain and testes. An adverse side effect of calcium EDTA is renal toxicity. Succimer (DMSA) is the preferred agent in mild to moderate lead poisoning cases. This may be the case in instances where children have a blood lead level >25µg/dL. The most reported adverse side effect for succimer is gastrointestinal disturbances. It is also important to note that chelation therapy only lowers blood lead levels and may not prevent the lead-induced cognitive problems associated with lower lead levels in tissue. This may be because of the inability of these agents to remove sufficient amounts of lead from tissue or inability to reverse preexisting damage. Chelating agents can have adverse effects; for example, chelation therapy can lower the body's levels of necessary nutrients like zinc. Chelating agents taken orally can increase the body's absorption of lead through the intestine.

Chelation challenge, also known as provocation testing, is used to indicate an elevated and mobilizable body burden of heavy metals including lead. This testing involves collecting urine before and after administering a one-off dose of chelating agent to mobilize heavy metals into the urine.

Then urine is analyzed by a laboratory for levels of heavy metals; from this analysis overall body burden is inferred. Chelation challenge mainly measures the burden of lead in soft tissues, though whether it accurately reflects long-term exposure or the amount of lead stored in bone remains controversial. Although the technique has been used to determine whether chelation therapy is indicated and to diagnose heavy metal exposure, some evidence does not support these uses as blood levels after chelation are not comparable to the reference range typically used to diagnose heavy metal poisoning. The single chelation dose could also redistribute the heavy metals to more sensitive areas such as central nervous system tissue.

Epidemiology

Since lead has been used widely for centuries, the effects of exposure are worldwide. Environmental lead is ubiquitous, and everyone has some measurable blood lead level. Atmospheric lead pollution increased dramatically beginning in the 1950s as a result of the widespread use of leaded gasoline. Lead is one of the largest environmental medicine problems in terms of numbers of people exposed and the public health toll it takes. Lead exposure accounts for about 0.2% of all deaths and 0.6% of disability adjusted life years globally.

Although regulation reducing lead in products has greatly reduced exposure in the developed world since the 1970s, lead is still allowed in products in many developing countries. In all countries that have banned leaded gasoline, average blood lead levels have fallen sharply. However, some developing countries still allow leaded gasoline, which is the primary source of lead exposure in most developing countries. Beyond exposure from gasoline, the frequent use of pesticides in developing countries adds a risk of lead exposure and subsequent poisoning. Poor children in developing countries are at especially high risk for lead poisoning. Of North American children, 7% have blood lead levels above 10 µg/dL, whereas among Central and South American children, the percentage is 33 to 34%. About one fifth of the world's disease burden from lead poisoning occurs in the Western Pacific, and another fifth is in Southeast Asia.

In developed countries, people with low levels of education living in poorer areas are most at risk for elevated lead. In the US, the groups most at risk for lead exposure are the impoverished, city-dwellers, and immigrants. African-American children and those living in old housing have also been found to be at elevated risk for high blood lead levels in the US. Low-income people often live in old housing with lead paint, which may begin to peel, exposing residents to high levels of lead-containing dust.

Risk factors for elevated lead exposure include alcohol consumption and smoking (possibly because of contamination of tobacco leaves with lead-containing pesticides). Adults with certain risk factors might be more susceptible to toxicity; these include calcium and iron deficiencies, old age, disease of organs targeted by lead (e.g. the brain, the kidneys), and possibly genetic susceptibility. Differences in vulnerability to lead-induced neurological damage between males and females have also been found, but some studies have found males to be at greater risk, while others have found females to be.

In adults, blood lead levels steadily increase with increasing age. In adults of all ages, men have higher blood lead levels than women do. Children are more sensitive to elevated blood lead levels than adults are. Children may also have a higher intake of lead than adults; they breathe faster and may be more likely to have contact with and ingest soil. Children of ages one to three tend to have the highest blood lead levels, possibly because at that age they begin to walk and explore their environment, and they use their mouths in their exploration. Blood levels usually peak at about 18–24 months old. In many countries including the US, household paint and dust are the major route of exposure in children.

Notable Cases

Cases of mass lead poisoning can occur, primarily in underdeveloped countries.

15,000 people are being relocated from Jiyuan in central Henan province to other locations after 1000 children living around China's largest smelter plant (owned and operated by Yuguang Gold and Lead) were found to have excess lead in their blood. The total cost of this project is estimated to around 1 billion yuan ($150 million). 70% of the cost will be paid by local government and the smelter company, while the rest will be paid by the residents themselves. The government has suspended production at 32 of 35 lead plants. The affected area includes people from 10 different villages.

The Zamfara State lead poisoning epidemic occurred in Nigeria in 2010. As of October 5, 2010 at least 400 children have died from the effects of lead poisoning.

Prognosis

Reversibility

Outcome is related to the extent and duration of lead exposure. Effects of lead on the physiology of the kidneys and blood are generally reversible; its effects on the central nervous system are not. While peripheral effects in adults often go away when lead exposure ceases, evidence suggests that most of lead's effects on a child's central nervous system are irreversible. Children with lead poisoning may thus have adverse health, cognitive, and behavioral effects that follow them into adulthood.

Encephalopathy

Lead encephalopathy is a medical emergency and causes permanent brain damage in 70–80% of children affected by it, even those that receive the best treatment. The mortality rate for people who develop cerebral involvement is about 25%, and of those who survive who had lead encephalopathy symptoms by the time chelation therapy was begun, about 40% have permanent neurological problems such as cerebral palsy.

Long-term

Exposure to lead may also decrease lifespan and have health effects in the long term. Death rates from a variety of causes have been found to be higher in people with elevated blood lead levels; these include cancer, stroke, and heart disease, and general death rates from all causes. Lead is considered a possible human carcinogen based on evidence from animal studies. Evidence also suggests that age-related mental decline and psychiatric symptoms are correlated with lead exposure. Cumulative exposure over a prolonged period may have a more important effect on some aspects of health than recent exposure. Some health effects, such as high blood pressure, are only significant risks when lead exposure is prolonged (over about one year).

History

Dioscorides noted lead's effect on the mind in the first century A.D.

Roman lead water pipes with taps

Lead poisoning was among the first known and most widely studied work and environmental hazards. One of the first metals to be smelted and used, lead is thought to have been discovered and first mined in Anatolia around 6500 BCE. Its density, workability, and corrosion-resistance were among the metal's attractions.

In the 2nd century BCE the Greek botanist Nicander described the colic and paralysis seen in lead-poisoned people. Dioscorides, a Greek physician who lived in the 1st century CE, wrote that lead makes the mind "give way".

Lead was used extensively in Roman aqueducts from about 500 BCE to 300 CE Julius Caesar's engineer, Vitruvius, reported, "water is much more wholesome from earthenware pipes than from lead pipes. For it seems to be made injurious by lead, because white lead is produced by it, and this is said to be harmful to the human body." Gout, prevalent in affluent Rome, is thought to be the result of lead, or leaded eating and drinking vessels. Sugar of lead (lead(II) acetate) was used to sweeten wine, and the gout that resulted from this was known as "saturnine" gout. It is even hypothesized that lead poisoning may have contributed to the decline of the Roman Empire, a hypothesis thoroughly disputed:

The great disadvantage of lead has always been that it is poisonous. This was fully recognised by the ancients, and Vitruvius specifically warns against its use. Because it was nevertheless used in profusion for carrying drinking water, the conclusion has often been drawn that the Romans must therefore have suffered from lead poisoning; sometimes conclusions are carried even further and it is inferred that this caused infertility and other unwelcome conditions, and that lead plumbing was largely responsible for the decline and fall of Rome.

Two things make this otherwise attractive hypothesis impossible. First, the calcium carbonate deposit that formed so thickly inside the aqueduct channels also formed inside the pipes, effectively insulating the water from the lead, so that the two never touched. Second, because the Romans had so few taps and the water was constantly running, it was never inside the pipes for more than a few minutes, and certainly not long enough to become contaminated.

However, recent research supports the idea that the lead found in the water came from the supply pipes, rather than another source of contamination. It was not unknown for locals to punch holes in the pipes to draw water off, increasing the number of people exposed to the lead.

Thirty years ago, Jerome Nriagu argued in a milestone paper that Roman civilization collapsed as a result of lead poisoning. Clair Patterson, the scientist who convinced governments to ban lead from gasoline, enthusiastically endorsed this idea, which nevertheless triggered a volley of publications aimed at refuting it. Although today lead is no longer seen as the prime culprit of Rome's demise, its status in the system of water distribution by lead pipes (fistulæ) still stands as a major public health issue. By measuring Pb isotope compositions of sediments from the Tiber River and the Trajanic Harbor, the present work shows that "tap water" from ancient Rome had 100 times more lead than local spring waters.

Romans also consumed lead through the consumption of defrutum, carenum, and sapa, musts made by boiling down fruit in lead cookware. Defrutum and its relatives were used in Ancient Roman cuisine and cosmetics, including as a food preservative. The use of leaden cookware, though popular, was not the general standard and copper cookware was used far more generally. There is also no indication how often sapa was added or in what quantity.

The consumption of sapa as having a role in the fall of the Roman Empire was used in a theory proposed by geochemist Jerome Nriagu to state that "lead poisoning contributed to the decline of the Roman Empire". In 1984, John Scarborough, a pharmacologist and classicist, criticized the conclusions drawn by Nriagu's book as "so full of false evidence, miscitations, typographical errors, and a blatant flippancy regarding primary sources that the reader cannot trust the basic arguments."

After antiquity, mention of lead poisoning was absent from medical literature until the end of the Middle Ages. In 1656 the German physician Samuel Stockhausen recognized dust and fumes containing lead compounds as the cause of disease, called since ancient Roman times *morbi metallici*, that were known to afflict miners, smelter workers, potters, and others whose work exposed them to the metal.

The painter Caravaggio might have died of lead poisoning. Bones with high lead levels were recently found in a grave thought likely to be his. Paints used at the time contained high amounts of lead salts. Caravaggio is known to have exhibited violent behavior, as caused by lead poisoning.

In 17th-century Germany, the physician Eberhard Gockel discovered lead-contaminated wine to be the cause of an epidemic of colic. He had noticed that monks who did not drink wine were healthy, while wine drinkers developed colic, and traced the cause to sugar of lead, made by simmering litharge with vinegar. As a result, Eberhard Ludwig, Duke of Württemberg issued an edict in 1696 banning the adulteration of wines with litharge.

In the 18th century lead poisoning was fairly frequent on account of the widespread drinking of rum, which was made in stills with a lead component (the "worm"). It was a significant cause of mortality amongst slaves and sailors in the colonial West Indies. Lead poisoning from rum was also noted in Boston. Benjamin Franklin suspected lead to be a risk in 1786. Also in the 18th century, "Devonshire colic" was the name given to the symptoms suffered by people of Devon who drank cider made in presses that were lined with lead. Lead was added to cheap wine illegally in the 18th and early 19th centuries as a sweetener. The composer Beethoven, a heavy wine drinker, suffered elevated lead levels (as later detected in his hair) possibly due to this; the cause of his death is controversial, but lead poisoning is a contender as a factor.

With the Industrial Revolution in the 19th century, lead poisoning became common in the work setting. The introduction of lead paint for residential use in the 19th century increased childhood exposure to lead; for millennia before this, most lead exposure had been occupational. An important step in the understanding of childhood lead poisoning occurred when toxicity in children from lead paint was recognized in Australia in 1897. France, Belgium, and Austria banned white lead interior paints in 1909; the League of Nations followed suit in 1922. However, in the United States, laws banning lead house paint were not passed until 1971, and it was phased out and not fully banned until 1978.

The 20th century saw an increase in worldwide lead exposure levels due to the increased widespread use of the metal. Beginning in the 1920s, lead was added to gasoline to improve its combustion; lead from this exhaust persists today in soil and dust in buildings. Blood lead levels worldwide have been declining sharply since the 1980s, when leaded gasoline began to be phased out. In those countries that have banned lead in solder for food and drink cans and have banned leaded gasoline additives, blood lead levels have fallen sharply since the mid-1980s.

The levels found today in most people are orders of magnitude greater than those of pre-industrial society. Due to reductions of lead in products and the workplace, acute lead poisoning is rare in most countries today, but low level lead exposure is still common. It was not until the second half of the 20th century that subclinical lead exposure became understood to be a problem. During the end of the 20th century, the blood lead levels deemed acceptable steadily declined. Blood lead levels once considered safe are now considered hazardous, with no known safe threshold.

In the 1980s Herbert Needleman was falsely accused of scientific misconduct by the Lead Industry Associates.

In 2002 Tommy Thompson, secretary of Health and Human Services appointed at least two persons with conflicts of interest to the CDC's Lead Advisory Committee.

In 2014 a case by the state of California against a number of companies decided against Sherwin-Williams NL Industries and ConAgra and ordered them to pay $1.15 billion. The case is currently under appeal.

Studies have found a weak link between lead from leaded gasoline and crime rates.

Other Species

Humans are not alone in suffering from lead's effects; plants and animals are also affected by lead toxicity to varying degrees depending on species. Animals experience many of the same effects of lead exposure as humans do, such as abdominal pain, peripheral neuropathy, and behavioral changes such as increased aggression. Much of what is known about human lead toxicity and its effects is derived from animal studies. Animals are used to test the effects of treatments, such as chelating agents, and to provide information on the pathophysiology of lead, such as how it is absorbed and distributed in the body.

Farm animals such as cows and horses as well as pet animals are also susceptible to the effects of lead toxicity. Sources of lead exposure in pets can be the same as those that present health threats to humans sharing the environment, such as paint and blinds, and there is sometimes lead in toys made for pets. Lead poisoning in a pet dog may indicate that children in the same household are at increased risk for elevated lead levels.

Wildlife

Turkey vultures, *Cathartes aura* (shown), and California condors can be poisoned when they eat carcasses of animals shot with lead pellets.

Lead, one of the leading causes of toxicity in waterfowl, has been known to cause die-offs of wild bird populations. When hunters use lead shot, waterfowl such as ducks can ingest the spent pellets later and be poisoned; predators that eat these birds are also at risk. Lead shot-related waterfowl poisonings were first documented in the US in the 1880s. By 1919, the spent lead pellets from waterfowl hunting was positively identified as the source of waterfowl deaths. Lead shot has been banned for hunting waterfowl in several countries, including the US in 1991 and Canada in 1997. Other threats to wildlife include lead paint, sediment from lead mines and smelters, and lead weights from fishing lines. Lead in some fishing gear has been banned in several countries.

The critically endangered California condor has also been affected by lead poisoning. As scavengers, condors eat carcasses of game that have been shot but not retrieved, and with them the fragments from lead bullets; this increases their lead levels. Among condors around the Grand Canyon, lead poisoning due to eating lead shot is the most frequently diagnosed cause of death. In an effort to protect this species, in areas designated as the California condor's range the use of projectiles containing lead has been banned to hunt deer, feral pigs, elk, pronghorn antelope, coyotes, ground squirrels, and other non-game wildlife. Also, conservation programs exist which routinely capture condors, check their blood lead levels, and treat cases of poisoning.

Cadmium Poisoning

Cadmium is an extremely toxic metal commonly found in industrial workplaces. Due to its low permissible exposure limit, overexposures may occur even in situations where trace quantities of cadmium are found. Cadmium is used extensively in electroplating, although the nature of the operation does not generally lead to overexposures. Cadmium is also found in some industrial paints and may represent a hazard when sprayed. Operations involving removal of cadmium paints by scraping or blasting may pose a significant hazard. Cadmium is also present in the manufacturing of some types of batteries. Exposures to cadmium are addressed in specific standards for the general industry, shipyard employment, construction industry, and the agricultural industry.

Sources of Exposure

In the 1950s and 1960s industrial exposure to cadmium was high, but as the toxic effects of cadmium became apparent, industrial limits on cadmium exposure have been reduced in most industrialized nations and many policy makers agree on the need to reduce exposure further. While working with cadmium it is important to do so under a fume hood to protect against dangerous fumes. Brazing fillers which contain cadmium should be handled with care. Serious toxicity problems have resulted from long-term exposure to cadmium plating baths.

Buildup of cadmium levels in the water, air, and soil has been occurring particularly in industrial areas. Environmental exposure to cadmium has been particularly problematic in Japan where many people have consumed rice that was grown in cadmium-contaminated irrigation water. This phenomenon is known under the name itai-itai disease.

Food is another source of cadmium. Plants may only contain small or moderate amounts in non-industrial areas, but high levels may be found in the liver and kidneys of adult animals. The daily intake of cadmium through food varies by geographic region. Intake is reported to be approximately 8 to 30μg in Europe and the United States versus 59 to 113 μg in various areas of Japan.

Cigarettes are also a significant source of cadmium exposure. Although there is generally less cadmium in tobacco than in food, the lungs absorb cadmium more efficiently than the stomach.

Aside from tobacco smokers, people who live near hazardous waste sites or factories that release cadmium into the air have the potential for exposure to cadmium in air. However, numerous state and federal regulations in the United States control the amount of cadmium that can be released to the air from waste sites and incinerators so that properly regulated sites are not hazardous. The general population and people living near hazardous waste sites may be exposed to cadmium in contaminated food, dust, or water from unregulated releases or accidental releases. Numerous regulations and use of pollution controls are enforced to prevent such releases.

Workers can be exposed to cadmium in air from the smelting and refining of metals, or from the air in plants that make cadmium products such as batteries, coatings, or plastics. Workers can also be exposed when soldering or welding metal that contains cadmium. Approximately 512,000 workers in the United States are in environments each year where a cadmium exposure may occur. Regulations that set permissible levels of exposure, however, are enforced to protect workers and to make sure that levels of cadmium in the air are considerably below levels thought to result in harmful effects.

Artists who work with cadmium pigments, which are commonly used in strong oranges, reds, and yellows, can easily accidentally ingest dangerous amounts, particularly if they use the pigments in dry form, as with chalk pastels, or in mixing their own paints.

Some sources of phosphate in fertilizers contain cadmium in amounts of up to 100 mg/kg, which can lead to an increase in the concentration of cadmium in soil (for example in New Zealand). Nickel-cadmium batteries are one of the most popular and most common cadmium-based products, and this soil can be mined for use in them.

An experiment during the early 1960s involving the spraying of cadmium over Norwich has recently been declassified by the UK government, as documented in a BBC News article.

In February 2010, cadmium was found in an entire line of Wal-Mart exclusive Miley Cyrus jewelry. The charms were tested at the behest of the *Associated Press* and were found to contain high levels of cadmium. Wal-Mart did not stop selling the jewelry until May 12 because "it would be too difficult to test products already on its shelves". On June 4 cadmium was detected in the paint used on promotional drinking glasses for the movie *Shrek Forever After*, sold by McDonald's Restaurants, triggering a recall of 12 million glasses.

Clinical Effects

Cadmium (Cd) is an extremely toxic industrial and environmental pollutant classified as a human carcinogen [Group 1 – according to International Agency for Research on Cancer; Group 2a – according to Environmental Protection Agency (EPA); and 1B carcinogen classified by European Chemical Agency

Acute exposure to cadmium fumes may cause flu-like symptoms including chills, fever, and muscle ache sometimes referred to as "the cadmium blues." Symptoms may resolve after a week if there is no respiratory damage. More severe exposures can cause tracheo-bronchitis, pneumonitis, and pulmonary edema. Symptoms of inflammation may start hours after the exposure and include cough, dryness and irritation of the nose and throat, headache, dizziness, weakness, fever, chills, and chest pain.

Inhaling cadmium-laden dust quickly leads to respiratory tract and kidney problems which can be fatal (often from renal failure). Ingestion of any significant amount of cadmium causes immediate poisoning and damage to the liver and the kidneys. Compounds containing cadmium are also carcinogenic.

The bones become soft (*osteomalacia*), lose bone mineral density (*osteoporosis*) and become weaker. This causes the pain in the joints and the back, and also increases the risk of fractures. In extreme cases of cadmium poisoning, mere body weight causes a fracture.

The kidneys lose their function to remove acids from the blood in *proximal renal tubular dysfunction*. The kidney damage inflicted by cadmium poisoning is irreversible. The *proximal renal tubular dysfunction* creates low phosphate levels in the blood (*hypophosphatemia*), causing muscle weakness and sometimes coma. The dysfunction also causes gout, a form of arthritis due to the accumulation of uric acid crystals in the joints because of high acidity of the blood (*hyperuricemia*). Another side effect is increased levels of chloride in the blood (*hyperchloremia*). The kidneys can also shrink up to 30%. Cadmium exposure is also associated with the development of kidney stones.

Other patients lose their sense of smell (anosmia).

Biological or Biochemical Mechanisms for Toxicity

Cadmium acts as a catalyst in forming reactive oxygen species. It increases lipid peroxidation and additionally depletes antioxidants, glutathione and protein-bound sulfhydryl groups. It also promotes the production of inflammatory cytokines.

Biomarkers of Excessive Exposure

Increased concentrations of urinary beta-2 microglobulin can be an early indicator of renal dysfunction in persons chronically exposed to low but excessive levels of environmental cadmium. The urinary beta-2 microglobulin test is an indirect method of measuring cadmium exposure. Under some circumstances, the Occupational Health and Safety Administration requires screening for renal damage in workers with long-term exposure to high levels of cadmium. Blood or urine cadmium concentrations provide a better index of excessive exposure in industrial situations or following acute poisoning, whereas organ tissue (lung, liver, kidney) cadmium concentrations may be useful in fatalities resulting from either acute or chronic poisoning. Cadmium concentrations in healthy persons without excessive cadmium exposure are generally less than 1 µg/L in either blood or urine. The ACGIH biological exposure indices for blood and urine cadmium levels are 5 µg/L and 5 µg/g creatinine, respectively, in random specimens. Persons who have sustained renal damage due to chronic cadmium exposure often have blood or urine cadmium levels in a range of 25-50 µg/L or 25-75 µg/g creatinine, respectively. These ranges are usually 1000-3000 µg/L and 100-400 µg/g, respectively, in survivors of acute poisoning and may be substantially higher in fatal cases.

Oil Pollution Toxicity to Marine Fish

Oil pollution toxicity to marine fish has been observed from oil spills such as the *Exxon Valdez* disaster, and from nonpoint sources, such as runoff, which is the largest source of oil pollution in marine waters. Crude oil entering waterways from spills or runoff contain polycyclic aromatic hydrocarbons (PAHs), the most toxic components of oil. The route of PAH uptake into fish depends on many environmental factors and the properties of the PAH. The common routes are ingestion, ventilation of the gills, and dermal uptake. Fish exposed to these PAHs exhibit an array of toxic effects including genetic damage, morphological deformities, altered growth and development, decreased body size, inhibited swimming abilities and mortality. The morphological deformities of PAH exposure, such as fin and jaw malformations, result in significantly reduced survival in fish due to the reduction of swimming and feeding abilities. While the exact mechanism of PAH toxicity is unknown, there are four proposed mechanisms. The difficulty in finding a specific toxic mechanism is largely due to the wide variety of PAH compounds with differing properties.

History

Research on the environmental impact of the petroleum industry began in earnest, during the mid to late 20th century, as the oil industry developed and expanded. Large scale transport of crude oil increased as a result of the increasing worldwide demand for oil, subsequently increasing the number of oil spills. Oil spills provided perfect opportunities for scientists to examine the in situ effects of crude oil exposure to marine ecosystems, and collaborative efforts between the National Oceanic and Atmospheric Administration (NOAA) and the United States Coast Guard resulted in improved response efforts and detailed research on oil pollution's effects. The Exxon Valdez oil spill in 1989, and the Deepwater Horizon oil spill in 2010, both resulted in increased scientific knowledge on the specific effects of oil pollution toxicity to marine fish.

Exxon Valdez Oil Spill

Focused research on oil pollution toxicity to fish began in earnest in 1989, after the *Exxon Valdez* tanker struck a reef in Prince William Sound, Alaska and spilled approximately 11 million gallons of crude oil into the surrounding water. At the time, the Exxon Valdez oil spill was the largest in the history of the United States. There were many adverse ecological impacts of the spill including the loss of the loss of billions of Pacific herring and pink salmon eggs. Pacific herring were just beginning to spawn in late March when the spill occurred, resulting in nearly half of the populations eggs being exposed to crude oil. Pacific herring spawn in the intertidal and subtidal zones, making the vulnerable eggs easily exposed to pollution.

Deepwater Horizon Oil Spill

After April 20, 2010, when an explosion on the *Deepwater Horizon* Macondo oil drilling platform triggered the largest oil spill in US history, another opportunity for oil toxicity research was presented. Approximately 171 million gallons of crude oil flowed from the seafloor into the Gulf of Mexico, exposing the majority of surrounding biota. The Deepwater Horizon oil spill also coincided directly with spawning window of various ecologically and commercially important fish species, including yellowfin and Atlantic bluefin tuna. The oil spill directly affected Atlantic bluefin tuna, as approximately 12% of larval tuna were located in oil contaminated waters, and Gulf of Mexico is the only known spawning grounds for the western population of bluefin tuna.

Exposure to Oil

Oil spills, as well daily oil runoff from urbanized areas can lead to polycyclic aromatic hydrocarbon (PAHs) entering marine ecosystems. Once PAHs enter the marine environment, fish can be exposed to them via ingestion, ventilation of the gills, and dermal uptake. The major route of uptake will depend on the behavior of the species of fish and the physicochemical properties of the PAH of concern. Habitat can be a major deciding factor for the route of exposure. For example, demersal fish or fish that consume demersal fish are highly likely to ingest PAHs that have sorbed to the sediment, whereas fish that swim at the surface are at a higher risk for dermal exposure. Upon coming in contact with a PAH, bioavailability will effect how readily the PAH is taken up. The EPA identifies 16 major PAHs of concern and each of these PAHs has a different degree of bioavailability. For instance, PAHs with lower molecular weight are more bioavailable because they dissolve more readily in water and are therefore more bioavailable for fish within the water column. Similarly, hydrophilic PAHs are more bioavailable for uptake by fish. For this reason, usage of oil dispersants, like Corexit, to treat oil spills can increase the uptake of PAHs by increasing their solubility in water and making them more available for uptake via the gills. Once a PAH is taken up, the fish's metabolism can affect the duration and intensity of the exposure to target tissues. Fish are able to readily metabolize 99% of PAHs to a more hydrophilic metabolite through their hepato-bilary system. This allows for the excretion of PAHs. The rate of metabolism of PAHs will depend on the sex and size of the species. The ability to metabolize PAHs into a more hydrophillic form can prevent bioaccumulation and hault PAHs from being passed on to organisms further up the food web. Because oil can persist in the environment long after oil spills via sedimentation, demersal fish are likely to be continually exposed to PAHs many years after oil spills. This has been proven by looking at the biliary PAH metabolites of bottom dwelling fish. For instance, bottom dwelling fish still showed elevated levels of low molecular weight PAH metabolites 10 years after *Exxon Valdez* oil spill.

Crude Oil Components

Crude oil is composed of more than 17,000 compounds. Among these 17,000 compounds are PAHs, which are considered the most toxic components of oil. PAHs are formed by pyrogenic and petrogenic processes. Petrogenic PAHs are formed by the elevated pressure of organic material. In contrast, pyrogenic PAHs are formed through the incomplete combustion of organic material. Crude oil naturally contains petrogenic PAHs and these PAH levels are increased significantly through the burning of oil which creates pyrogenic PAHs. The level of PAHs found in crude oil differs with the type of crude oil. For example, crude oil from the *Exxon Valdez* oil spill had PAH concentrations of 1.47%, while PAH concentrations from the North Sea have much lower PAH concentrations of 0.83%.

Sources of Crude Oil Pollution

Crude oil contamination in marine ecosystems can lead to both pyrogenic and petrogenic PAHs entering these ecosystems. Petrogenic PAHs can enter waterways through oil seeps, major oil spills, creosote and fuel oil runoff from urban areas. Pyrogenic PAH sources consist of diesel soot tire rubber and coal dust. Although there are natural sources of PAHs such as volcanic activity and seepage of coal deposits, anthropogenic sources pose the most significant input of PAHs into the environment. These anthroprogenic sources include residential heating, asphalt production, coal gasification, and petroleum usage. Petrogenic PAH contamination is more common from crude oil spills such as *Exxon Valdez*, or oil seeps; however, with runoff pyrogenic PAHs can also be prevalent. Although major oil spills such as *Exxon Valdez* can introduce a large amount of crude oil to a localized area in a short time span, daily runoff comprises most of the oil pollution to marine ecosystems. Atmospheric deposition can also be a source of PAHs into marine ecosystems. The deposition of PAHs from the atmosphere into a water body is largely influenced by the gas-particle partitioning of the PAH.

Effects

Many effects of PAH exposure have been observed in marine fish. Specifically, studies have been conducted on the embryonic and larval fish, the development of fish exposed to PAHs, and uptake of PAHs by fish via various routes of exposure. One study on found that Pacific herring eggs exposed to conditions mimicking the "Exxon Valdez" oil spill resulted in premature hatching of eggs, reduced size as fish matured and significant teratogenic effects, including skeletal, cardiovascular, fin and yolk sac malformations. Yolk sac edema was responsible for the majority of herring larval mortality. The teratogenic malformations in the dorsal fin and spine, and in the jaw were observed to effectively decrease the survival of developing fish, through the impairing of swimming and feeding ability respectively. Feeding and prey avoidance via swimming are crucial for the survival of larval and juvenile fish. All effects observed in herring eggs in the study were consistent with effects observed in exposed fish eggs following the *Exxon Valdez* oil spill. Zebrafish embryos exposed to oil were observed to have severe teratogenic defects similar to those seen in herring embryos, including edema, cardiac dysfunction, and intracranial hemorrhages. In a study focused on the uptake of PAHs by fish, salmon embryos were exposed to crude oil in three various situations, including via effluent from oil coated gravel. PAH concentrations in embryos directly exposed to oil and those exposed to PAH effluent were not significantly different. PAH exposure was observed to lead to death, even when the PAHs were exposed to fish via effluent. From the results, it was determined that fish embryos near the *Exxon Valdez* spill in Prince William Sound that were not directly in contact with oil still may have accumulated lethal levels of PAHs. While many laboratory and natural studies have observed significant adverse effects of PAH exposure to fish, a lack of effects has also been observed for certain PAH compounds, which could be due to a lack of uptake during exposure to the compound.

Proposed Mechanism of Toxic Action

While it has been proven that different classes of PAHs act through distinct toxic mechanisms due to the variations in their molecular weight, ring arrangements, and water solubility properties, the specific mechanisms of PAH toxicity to fish and fish development are still unknown. The proposed mechanisms of toxicity of PAHs are toxicity through narcosis, interaction with the AhR pathway, alkyl phenanthrene toxicity, and additive toxicity by multiple mechanisms.

- The narcosis model was not able to accurately predict the outcome of PAH mixture exposure of herring and pink salmon, according to a study.

- The primary toxicity of these PAHs in fish embryos has been observed to be AhR independent, and their cardiac effects are not associated with AhR activation or Cytochrome P450, family 1, member A induction in the endocardium.

- The alkyl phenanthrene model has been studied by exposing herring and pink salmon to mixtures of PAHs in an attempt to better understand the toxicity mechanisms of PAHs.

The model was found to generally predict the outcomes of sublethal and lethal exposures. Oxidative stress and effects on cardiovascular morphogenesis are proposed mechanisms for alkyl phenanthrene toxicity. The specific pathway is unknown.

- Since PAHs contain many different variations of PAHs, the toxicity may be explained by using multiple mechanisms of action.

Nuclear Fallout

Nuclear fallout, or simply *fallout,* is the residual radioactive material propelled into the upper atmosphere following a nuclear blast or a nuclear reaction conducted in an unshielded facility, so called because it "falls out" of the sky after the explosion and the shock wave have passed. It commonly refers to the radioactive dust and ash created when a nuclear weapon explodes, but such dust can also originate from a damaged nuclear plant. Fallout may take the form of black rain (rain darkened by particulates).

This radioactive dust, consisting of material either directly vaporized by a nuclear blast or charged by exposure, is a highly dangerous kind of radioactive contamination.

Types of Fallout

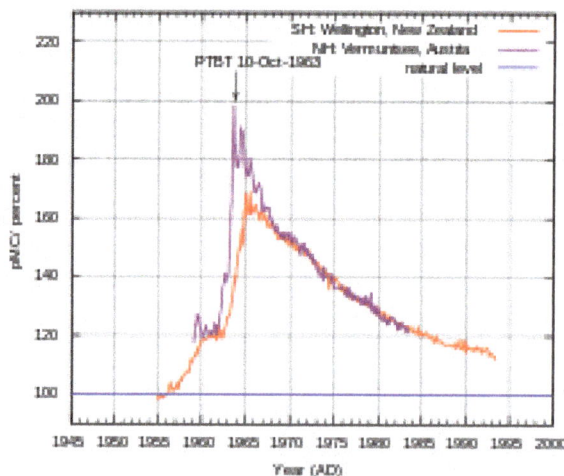

Atmospheric nuclear weapon tests almost doubled the concentration of radioactive ^{14}C in the Northern Hemisphere, before levels slowly declined following the Partial Test Ban Treaty.

An air burst (that is, a nuclear detonation far above the surface) can eventually produce worldwide fallout. A ground burst can produce possibly much more severe, local fallout.

Global Fallout

After an air burst, fission products, un-fissioned nuclear material, and weapon residues vaporized by the heat of the fireball condense into a fine suspension of small particles 10 nm to 20 μm in diameter. These particles may be quickly drawn up into the stratosphere, particularly if the explosive yield exceeds 10 kt.

Initially little was known about the dispersion of nuclear fallout on a global scale. The AEC assumed that fallout would be dispersed evenly across the globe by atmospheric winds and gradually settle to the Earth's surface after weeks, months, and even years as worldwide fallout.

The radio-biological hazard of worldwide fallout is a long-term one because of the potential accumulation of long-lived radioisotopes (such as strontium-90 and caesium-137) in the body as a result of ingestion of foods containing the radioactive materials. This hazard is less pertinent than local fallout, which is of much greater immediate operational concern.

Local fallout

In a land or water surface burst, heat vaporizes large amounts of earth or water, which is drawn up into the radioactive cloud. This material becomes radioactive when it condenses with fission products and other radiocontaminants that have become neutron-activated. The table below summarizes the abilities of common isotopes to form fallout. Some radiation taints large amounts of land and drinking water causing formal mutations throughout animal and human life.

The 450 km (280 mi) fallout plume from 15 Mt shot Castle Bravo, 1954

Table (according to T. Imanaka et al.) of the relative abilities of isotopes to form solids											
Isotope	^{91}Sr	^{92}Sr	^{95}Zr	^{99}Mo	^{106}Ru	^{131}Sb	^{132}Te	^{134}Te	^{137}Cs	^{140}Ba	^{141}La	^{144}Ce
Refractory index	0.2	1.0	1.0	1.0	0.0	0.1	0.0	0.0	0.0	0.3	0.7	1.0

Per capita thyroid doses in the continental United States resulting from all exposure routes from all atmospheric nuclear tests conducted at the Nevada Test Site from 1951-1962.

A surface burst generates large amounts of particulate matter, composed of particles from less than 100 nm to several millimeters in diameter—in addition to very fine particles that contribute to worldwide fallout. The larger particles spill out of the stem and cascade down the outside of the fireball in a downdraft even as the cloud rises, so fallout begins to arrive near ground zero within an hour. More than half the total bomb debris lands on the ground within about 24 hours as local fallout. Chemical properties of the elements in the fallout control the rate at which they are deposited on the ground. Less volatile elements deposit first.

Severe local fallout contamination can extend far beyond the blast and thermal effects, particularly in the case of high yield surface detonations. The ground track of fallout from an explosion depends on the weather from the time of detonation onwards. In stronger winds, fallout travels faster but takes the same time to descend, so although it covers a larger path, it is more spread out or diluted. Thus, the width of the fallout pattern for any given dose rate is reduced where the downwind distance is increased by higher winds. The total amount of activity deposited up to any given time is the same irrespective of the wind pattern, so overall casualty figures from fallout are generally independent of winds. But thunderstorms can bring down activity as rain more rapidly than dry fallout, particularly if the mushroom cloud is low enough to be below ("washout"), or mixed with ("rainout"), the thunderstorm.

Whenever individuals remain in a radiologically contaminated area, such contamination leads to an immediate external radiation exposure as well as a possible later internal hazard from inhalation and ingestion of radiocontaminants, such as the rather short-lived iodine-131, which is accumulated in the thyroid.

Factors Affecting Fallout

Location

There are two main considerations for the location of an explosion: height and surface composition. A nuclear weapon detonated in the air, called an air burst, produces less fallout than a comparable explosion near the ground.

In case of water surface bursts, the particles tend to be rather lighter and smaller, producing less local fallout but extending over a greater area. The particles contain mostly sea salts with some water; these can have a cloud seeding effect causing local rainout and areas of high local fallout. Fallout from a seawater burst is difficult to remove once it has soaked into porous surfaces because the fission products are present as metallic ions that chemically bond to many surfaces. Water and detergent washing effectively removes less than 50% of this chemically bonded activity from concrete or steel. Complete decontamination requires aggressive treatment like sandblasting, or acidic treatment. After the *Crossroads* underwater test, it was found that wet fallout must be immediately removed from ships by continuous water washdown (such as from the fire sprinkler system on the decks).

Parts of the sea bottom may become fallout. After the Castle Bravo test, white dust—contaminated calcium oxide particles originating from pulverized and calcined corals—fell for several hours, causing beta burns and radiation exposure to the inhabitants of the nearby atolls and the crew of the Daigo Fukuryū Maru fishing boat. The scientists called the fallout Bikini snow.

For subsurface bursts, there is an additional phenomenon present called "base surge". The base surge is a cloud that rolls outward from the bottom of the subsiding column, which is caused by an excessive density of dust or water droplets in the air. For underwater bursts, the visible surge is, in effect, a cloud of liquid (usually water) droplets with the property of flowing almost as if it were a homogeneous fluid. After the water evaporates, an invisible base surge of small radioactive particles may persist.

For subsurface land bursts, the surge is made up of small solid particles, but it still behaves like a fluid. A soil earth medium favors base surge formation in an underground burst. Although the base surge typically contains only about 10% of the total bomb debris in a subsurface burst, it can create larger radiation doses than fallout near the detonation, because it arrives sooner than fallout, before much radioactive decay has occurred.

Meteorological

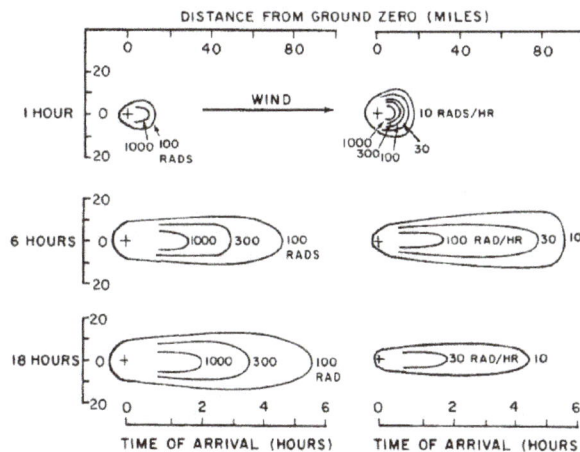

Comparison of fallout gamma dose and dose rate contours for a 1 Mt fission land surface burst, based on DELFIC calculations. Because of radioactive decay, the dose rate contours contract after fallout has arrived, but dose contours continue to grow

Meteorological conditions greatly influence fallout, particularly local fallout. Atmospheric winds are able to bring fallout over large areas. For example, as a result of a *Castle Bravo* surface burst of a 15 Mt thermonuclear device at Bikini Atoll on March 1, 1954, a roughly cigar-shaped area of the Pacific extending over 500 km downwind and varying in width to a maximum of 100 km was severely contaminated. There are three very different versions of the fallout pattern from this test, because the fallout was only measured on a small number of widely spaced Pacific Atolls. The two alternative versions both ascribe the high radiation levels at north Rongelap to a downwind hotspot caused by the large amount of radioactivity carried on fallout particles of about 50-100 micrometres size.

After *Bravo*, it was discovered that fallout landing on the ocean disperses in the top water layer (above the thermocline at 100 m depth), and the land equivalent dose rate can be calculated by multiplying the ocean dose rate at two days after burst by a factor of about 530. In other 1954 tests, including *Yankee* and *Nectar,* hotspots were mapped out by ships with submersible probes, and similar hotspots occurred in 1956 tests such as *Zuni* and *Tewa.* However, the major U.S. 'DELFIC' (Defence Land Fallout Interpretive Code) computer calculations use the natural size distributions of particles in soil instead of the afterwind sweep-up spectrum, and this results in more straight-forward fallout patterns lacking the downwind hotspot.

Snow and rain, especially if they come from considerable heights, accelerate local fallout. Under special meteorological conditions, such as a local rain shower that originates above the radioactive cloud, limited areas of heavy contamination just downwind of a nuclear blast may be formed.

Effects

A wide range of biological changes may follow the irradiation of animals. These vary from rapid death following high doses of penetrating whole-body radiation, to essentially normal lives for a variable period of time until the development of delayed radiation effects, in a portion of the exposed population, following low dose exposures.

The unit of actual *exposure* is the röntgen, defined in ionisations per unit volume of air. All ionisation based instruments (including geiger counters and ionisation chambers) measure exposure. However, effects depend on the energy per unit mass, not the exposure measured in air. A deposit of 1 joule per kilogram has the unit of 1 gray (Gy). For 1 MeV energy gamma rays, an exposure of 1 röntgen in air produces a dose of about 0.01 gray (1 centigray, cGy) in water or surface tissue. Because of shielding by the tissue surrounding the bones, the bone marrow only receives about 0.67 cGy when the air exposure is 1 röntgen and the surface skin dose is 1 cGy. Some lower values reported for the amount of radiation that would kill 50% of personnel (the LD_{50}) refer to bone marrow dose, which is only 67% of the air dose.

Short Term

Fallout shelter sign on a building in New York City.

The dose that would be lethal to 50% of a population is a common parameter used to compare the effects of various fallout types or circumstances. Usually, the term is defined for a specific time, and limited to studies of acute lethality. The common time periods used are 30 days or less for most small laboratory animals and to 60 days for large animals and humans. The LD_{50} figure assumes that the individuals did not receive other injuries or medical treatment.

In the 1950s, the LD_{50} for gamma rays was set at 3.5 Gy, while under more dire conditions of war (a bad diet, little medical care, poor nursing) the LD_{50} was 2.5 Gy (250 rad). There have been few documented cases of survival beyond 6 Gy. One person at Chernobyl survived a dose of more than 10 Gy, but many of the persons exposed there were not uniformly exposed over their entire body. If a person is exposed in a non-homogeneous manner then a given dose (averaged over the entire body) is less likely to be lethal. For instance, if a person gets a hand/low arm dose of 100 Gy, which gives them an overall dose of 4 Gy, they are more likely to survive than a person who gets a 4 Gy dose over their entire body. A hand dose of 10 Gy or more would likely result in loss of the hand. A British industrial radiographer who was estimated to have received a hand dose of 100 Gy over the course of his lifetime lost his hand because of radiation dermatitis. Most people become ill after an exposure to 1 Gy or more. The fetuses of pregnant women are often more vulnerable to radiation and may miscarry, especially in the first trimester.

One hour after a surface burst, the radiation from fallout in the crater region is 30 grays per hour (Gy/h). Civilian dose rates in peacetime range from 30 to 100 μGy per year.

Fallout radiation decays exponentially relatively quickly with time. Most areas become fairly safe for travel and decontamination after three to five weeks.

For yields of up to 10 kt, prompt radiation is the dominant producer of casualties on the battlefield. Humans receiving an acute incapacitating dose (30 Gy) have their performance degraded almost immediately and become ineffective within several hours. However, they do not die until five to six days after exposure, assuming they do not receive any other injuries. Individuals receiving less than a total of 1.5 Gy are not incapacitated. People receiving doses greater than 1.5 Gy become disabled, and some eventually die.

A dose of 5.3 Gy to 8.3 Gy is considered lethal but not immediately incapacitating. Personnel exposed to this amount of radiation have their performance degraded in two to three hours, depending on how physically demanding the tasks they must perform are, and remain in this disabled state at least two days. However, at that point they experience a recovery period and can perform non-demanding tasks for about six days, after which they relapse for about four weeks. At this time they begin exhibiting symptoms of radiation poisoning of sufficient severity to render them totally ineffective. Death follows at approximately six weeks after exposure, although outcomes may vary.

Long Term

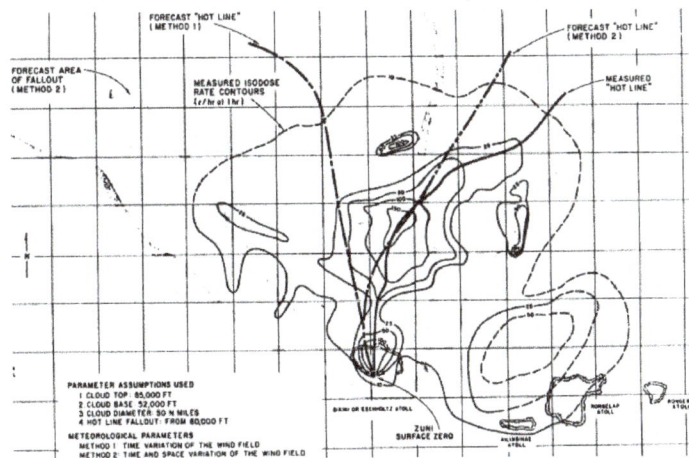

Comparison of predicted fallout "hotline" with test results in the 3.53 Mt 15% fission *Zuni* test at Bikini in 1956. The predictions were made under simulated tactical nuclear war conditions aboard ship by Edward A. Schuert.

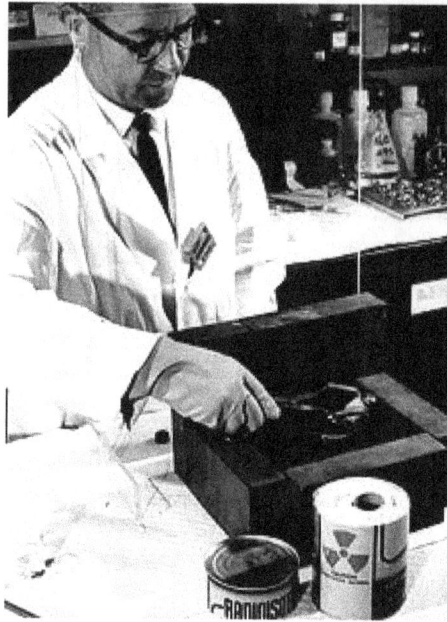

Following the detonation of the first atomic bomb, pre-war steel became a valuable commodity for scientists since it was the only steel not contaminated by radiation.

Late or delayed effects of radiation occur following a wide range of doses and dose rates. Delayed effects may appear months to years after irradiation and include a wide variety of effects involving almost all tissues or organs. Some of the possible delayed consequences of radiation injury are life shortening, carcinogenesis, cataract formation, chronic radiodermatitis, decreased fertility, and genetic mutations. Presently, the only teratological effect observed in humans following nuclear attacks on highly populated areas is microcephaly which is the only proven malformation, or congenital abnormality, found in the in utero developing human fetuses present during the Hiroshima and Nagasaki bombings. Of all the pregnant women exposed in the two cities, the number of children born with microcephaly was below 50. No statistically demonstrable increase of congenital malformations was found among the *later conceived children* born to survivors of the nuclear detonations at Hiroshima and Nagasaki. The surviving women of Hiroshima and Nagasaki who could conceive and were exposed to substantial amounts of radiation went on and had children with no higher incidence of abnormalities than the Japanese average.

The Baby Tooth Survey helped to determine the effects of nuclear fallout in the human anatomy by examining the levels of radioactive material absorbed into the deciduous teeth of children. Founded by the husband and wife team of physicians Eric Reiss and Louise Reiss, the research focused on detecting the presence of strontium-90, a cancer-causing radioactive isotope created by the more than 400 atomic tests conducted above ground that is absorbed from water and dairy products into the bones and teeth given its chemical similarity to calcium. The team sent collection forms to schools in the St. Louis, Missouri area, hoping to gather 50,000 teeth each year. Ultimately, the project collected over 300,000 teeth from children of various ages before the project was ended in 1970.

Preliminary results of the Baby Tooth Survey were published in the November 24, 1961, edition of the journal *Science*, and showed that levels of strontium 90 had risen steadily in children born in the 1950s, with those born later showing the most pronounced increases. The results of a more comprehensive study of the elements found in the teeth collected showed that children born after 1963 had levels of strontium 90 in their baby teeth that was 50 times higher than that found in children born before large-scale atomic testing began. The findings helped convince U.S. President John F. Kennedy to sign the Partial Nuclear Test Ban Treaty with the United Kingdom and Soviet Union, which ended the above-ground nuclear weapons testing that created the greatest amounts of atmospheric nuclear fallout.

Fallout Protection

During the Cold War, the governments of the U.S., the USSR, Great Britain, and China attempted to educate their citizens about surviving a nuclear attack by providing procedures on minimizing short-term exposure to fallout. This effort commonly became known as Civil Defense.

Fallout protection is almost exclusively concerned with protection from radiation. Radiation from fallout is encountered in the forms of alpha, beta, and gamma radiation, and as ordinary clothing affords protection from alpha and beta radiation, most fallout protection measures deal with reducing exposure to gamma radiation. For the purposes of radiation shielding, many materials have a characteristic *halving thickness*: the thickness of a layer of a material sufficient to reduce gamma radiation exposure by 50%. Halving thicknesses of common materials include: 1 cm (0.4 inch) of lead, 6 cm (2.4 inches) of concrete, 9 cm (3.6 inches) of packed earth or 150 m (500 ft) of air. When multiple thicknesses are built, the shielding multiplies. A practical fallout shield is ten halving-thicknesses of a given material, such as 90 cm (36 inches) of packed earth, which reduces gamma ray exposure by approximately 1024 times (2^{10}). A shelter built with these materials for the purposes of fallout protection is known as a fallout shelter.

The Seven Ten Rule

The danger of radiation from fallout also decreases with time, as radioactivity decays exponentially with time, such that for each factor of seven increase in time, the radiation is reduced by a factor of ten. For example, after 7 hours, the average dose rate is reduced by a factor of ten; after 49 hours, it is reduced by a further factor of ten (to 1/100th); after two weeks the radiation from the fallout will have reduced by a factor of 1000 compared the initial level; and after 14 weeks the average dose rate will have reduced to 1/10,000th of the initial level.

Nuclear Reactor Accident

Fallout can also refer to nuclear accidents, although a nuclear reactor does not explode like a nuclear weapon. The isotopic signature of bomb fallout is very different from the fallout from a serious power reactor accident (such as Chernobyl or Fukushima). The Fukushima plants have tons of nuclear fuel, thousands of Fuel Assemblies, more than 6,000 fuel rods in spent fuel pools.

The key differences are in volatility and half-life.

Volatility

The boiling point of an element (or its compounds) is able to control the percentage of that element a power reactor accident releases. The ability of an element to form a solid, controls the rate it is deposited on the ground after having been injected into the atmosphere by a nuclear detonation or accident.

Half-Life

A half life is the time it takes half of the radiation of a specific substance to decay. A large amount of short-lived isotopes such as ^{97}Zr are present in bomb fallout. This isotope and other short-lived isotopes are constantly generated in a power reactor, but because the criticality occurs over a long length of time, the majority of these short lived isotopes decay before they can be released.

DDT

DDT (dichlorodiphenyltrichloroethane) is a colorless, crystalline, tasteless and almost odorless organochlorine known for its insecticidal properties and environmental impacts. DDT has been formulated in multiple forms, including solutions in xylene or petroleum distillates, emulsifiable concentrates, water-wettable powders, granules, aerosols, smoke candles and charges for vaporizers and lotions.

First synthesized in 1874, DDT's insecticidal action was discovered by the Swiss chemist Paul Hermann Müller in 1939. It was used in the second half of World War II to control malaria and typhus among civilians and troops. After the war, DDT was also used as an agricultural insecticide and its production and use duly increased. Müller was awarded the Nobel Prize in Physiology or Medicine "for his discovery of the high efficiency of DDT as a contact poison against several arthropods" in 1948.

In 1962, Rachel Carson's book *Silent Spring* was published. It cataloged the environmental impacts of widespread DDT spraying in the United States and questioned the logic of releasing large amounts of potentially dangerous chemicals into the environment without understanding their effects on the environment or human health. The book claimed that DDT and other pesticides had been shown to cause cancer and that their agricultural use was a threat to wildlife, particularly birds. Its publication was a seminal event for the environmental movement and resulted in a large public outcry that eventually led, in 1972, to a ban on DDT's agricultural use in the United States. A worldwide ban on agricultural use was formalized under the Stockholm Convention on Persistent Organic Pollutants, but its limited and still-controversial use in disease vector control continues, because of its effectiveness in reducing malarial infections, balanced by environmental and other health concerns.

Along with the passage of the Endangered Species Act, the United States ban on DDT is cited by scientists as a major factor in the comeback of the bald eagle (the national bird of the United States) and the peregrine falcon from near-extinction in the contiguous United States.

Properties and Chemistry

DDT is similar in structure to the insecticide methoxychlor and the acaricide dicofol. It is highly hydrophobic and nearly insoluble in water but has good solubility in most organic solvents, fats and oils. DDT does not occur naturally. It is produced by the reaction of chloral (CCl_3CHO) with chlorobenzene (C

6H5Cl) in the presence of a sulfuric acid catalyst. DDT has been marketed under trade names including Anofex, Cezarex, Chlorophenothane, Clofenotane, Dicophane, Dinocide, Gesarol, Guesapon, Guesarol, Gyron, Ixodex, Neocid, Neocidol and Zerdane.

Isomers and Related Compounds

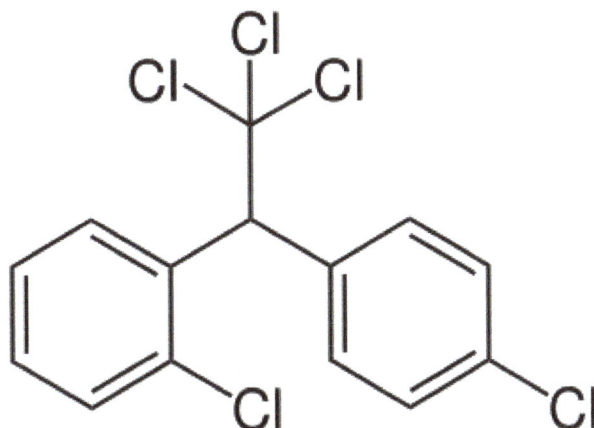

o,p' -DDT, a minor component in commercial DDT.

Commercial DDT is a mixture of several closely–related compounds. The major component (77%) is the *p,p'* isomer (pictured above). The *o,p'* isomer (pictured to the right) is also present in significant amounts (15%). Dichlorodiphenyldichloroethylene (DDE) and dichlorodiphenyldichloroethane (DDD) make up the balance. DDE and DDD are the major metabolites and environmental breakdown products. The term "total DDT" is often used to refer to the sum of all DDT related compounds (*p,p'*-DDT, *o,p'*-DDT, DDE, and DDD) in a sample.

Production and Use

From 1950 to 1980, DDT was extensively used in agriculture – more than 40,000 tonnes each year worldwide – and it has been estimated that a total of 1.8 million tonnes have been produced globally since the 1940s. In the United States, it was manufactured by some 15 companies, including Monsanto, Ciba, Montrose Chemical Company, Pennwalt and Velsicol Chemical Corporation. Production peaked in 1963 at 82,000 tonnes per year. More than 600,000 tonnes (1.35 billion pounds) were applied in the US before the 1972 ban. Usage peaked in 1959 at about 36,000 tonnes.

In 2009, 3,314 tonnes were produced for malaria control and visceral leishmaniasis. India is the only country still manufacturing DDT and is the largest consumer. China ceased production in 2007.

Mechanism of Insecticide Action

In insects it opens sodium ion channels in neurons, causing them to fire spontaneously, which leads to spasms and eventual death. Insects with certain mutations in their sodium channel gene are resistant to DDT and similar insecticides. DDT resistance is also conferred by up-regulation of genes expressing cytochrome P450 in some insect species, as greater quantities of some enzymes of this group accelerate the toxin's metabolism into inactive metabolites.

History

Commercial product concentrate containing 50% DDT, circa 1960s

Commercial product (Powder box, 50 g) containing 10% DDT; Néocide. Ciba Geigy DDT; *"Destroys parasites such as fleas, lice, ants, bedbugs, cockroaches, flies, etc.. Néocide Sprinkle caches of vermin and the places where there are insects and their places of passage. Leave the powder in place as long as possible." "Destroy the parasites of man and his dwelling". "Death is not instantaneous, it follows inevitably sooner or later." "French manufacturing"; "harmless to humans and warm-blooded animals" "sure and lasting effect. Odorless."*

External audio

"Episode 207: DDT", Chemical Heritage Foundation

DDT was first synthesized in 1874 by Othmar Zeidler under the supervision of Adolf von Baeyer. It was further described in 1929 in a dissertation by W. Bausch and in two subsequent publications in 1930. The insecticide properties of "multiple chlorinated aliphatic or fat-aromatic alcohols with at least one trichloromethane group" were described in a patent in 1934 by Wolfgang von Leuthold. DDT's insecticidal properties were not, however, discovered until 1939 by the Swiss scientist Paul Hermann Müller, who was awarded the 1948 Nobel Prize in Physiology and Medicine for his efforts.

Use in the 1940s and 1950s

DDT is the best-known of several chlorine-containing pesticides used in the 1940s and 1950s. With pyrethrum in short supply, DDT was used extensively during World War II by the Allies to control the insect vectors of typhus – nearly eliminating the disease in many parts of Europe. In the South Pacific, it was sprayed aerially for malaria and dengue fever control with spectacular effects. While DDT's chemical and insecticidal properties were important factors in these victories, advances in application equipment coupled with competent organization and sufficient manpower were also crucial to the success of these programs.

In 1945, DDT was made available to farmers as an agricultural insecticide and played a role in the final elimination of malaria in Europe and North America.

In 1955, the World Health Organization commenced a program to eradicate malaria in countries with low to moderate transmission rates worldwide, relying largely on DDT for mosquito control and rapid diagnosis and treatment to reduce transmission. The program eliminated the disease in "Taiwan, much of the Caribbean, the Balkans, parts of northern Africa, the northern region of Australia, and a large swath of the South Pacific" and dramatically reduced mortality in Sri Lanka and India.

However, failure to sustain the program, increasing mosquito tolerance to DDT, and increasing parasite tolerance led to a resurgence. In many areas early successes partially or completely reversed, and in some cases rates of transmission increased. The program succeeded in eliminating malaria only in areas with "high socio-economic status, well-organized healthcare systems, and relatively less intensive or seasonal malaria transmission".

DDT was less effective in tropical regions due to the continuous life cycle of mosquitoes and poor infrastructure. It was not applied at all in sub-Saharan Africa due to these perceived difficulties. Mortality rates in that area never declined to the same dramatic extent, and now constitute the bulk of malarial deaths worldwide, especially following the disease's resurgence as a result of resistance to drug treatments and the spread of the deadly malarial variant caused by *Plasmodium falciparum*.

Eradication was abandoned in 1969 and attention instead focused on controlling and treating the disease. Spraying programs (especially using DDT) were curtailed due to concerns over safety and environmental effects, as well as problems in administrative, managerial and financial implementation. Efforts shifted from spraying to the use of bednets impregnated with insecticides and other interventions.

United States Ban

As early as the 1940s, US scientists began expressing concern over possible hazards associated with DDT, and in the 1950s the government began tightening regulations governing its use. These events received little attention. In 1957 the *New York Times* reported an unsuccessful struggle to restrict DDT use in Nassau County, New York, that the issue came to the attention of the popular naturalist-author, Rachel Carson. William Shawn, editor of *The New Yorker*, urged her to write a piece on the subject, which developed into her 1962 book *Silent Spring*. The book argued that pesticides, including DDT, were poisoning both wildlife and the environment and were endangering human health. *Silent Spring* was a best seller, and public reaction to it launched the modern environmental movement in the United States. The year after it appeared, President John F. Kennedy ordered his Science Advisory Committee to investigate Carson's claims. The committee's report "add[ed] up to a fairly thorough-going vindication of Rachel Carson's Silent Spring thesis," in the words of the journal *Science*, and recommended a phaseout of "persistent toxic pesticides". DDT became a prime target of the growing anti-chemical and anti-pesticide movements, and in 1967 a group of scientists and lawyers founded the Environmental Defense Fund (EDF) with the specific goal of enacting a ban on DDT. Victor Yannacone, Charles Wurster, Art Cooley and others in the group had all witnessed bird kills or declines in bird populations and suspected that DDT was the cause. In their campaign against the chemical, EDF petitioned the government for a ban and filed lawsuits. Around this time, toxicologist David Peakall was measuring DDE levels in the eggs of peregrine falcons and California condors and finding that increased levels corresponded with thinner shells.

In response to an EDF suit, the U.S. District Court of Appeals in 1971 ordered the EPA to begin the de-registration procedure for DDT. After an initial six-month review process, William Ruckelshaus, the Agency's first Administrator rejected an immediate suspension of DDT's registration, citing studies from the EPA's internal staff stating that DDT was not an imminent danger. However, these findings were criticized, as they were performed mostly by economic entomologists inherited from the United States Department of Agriculture, who many environmentalists felt were biased towards agribusiness and understated concerns about human health and wildlife. The decision thus created controversy.

The EPA held seven months of hearings in 1971–1972, with scientists giving evidence for and against DDT. In the summer of 1972, Ruckelshaus announced the cancellation of most uses of DDT – exempting public health uses under some conditions. Immediately after the announcement, both EDF and the DDT manufacturers filed suit against EPA. Industry sought to overturn the ban, while EDF wanted a comprehensive ban. The cases were consolidated, and in 1973 the United States Court of Appeals for the District of Columbia Circuit ruled that the EPA had acted properly in banning DDT.

Some uses of DDT continued under the public health exemption. For example, in June 1979, the California Department of Health Services was permitted to use DDT to suppress flea vectors of bubonic plague. DDT continued to be produced in the United States for foreign markets until 1985, when over 300 tons were exported.

Restrictions on Usage

In the 1970s and 1980s, agricultural use was banned in most developed countries, beginning with Hungary in 1968 followed by Norway and Sweden in 1970, West Germany and the US in 1972, but not in the United Kingdom until 1984. By 1991 total bans, including for disease control, were in place in at least 26 countries; for example Cuba in 1970, Singapore in 1984, Chile in 1985 and the Republic of Korea in 1986.

The Stockholm Convention on Persistent Organic Pollutants, which took effect in 2004, outlawed several persistent organic pollutants, and restricted DDT use to vector control. The Convention was ratified by more than 170 countries. Recognizing that total elimination in many malaria-prone countries is currently unfeasible absent affordable/effective alternatives, the convention exempts public health use within World Health Organization (WHO) guidelines from the ban. Resolution 60.18 of the World Health Assembly commits WHO to the Stockholm Convention's aim of reducing and ultimately eliminating DDT. Malaria Foundation International states, "The outcome of the treaty is arguably better than the status quo going into the negotiations. For the first time, there is now an insecticide which is restricted to vector control only, meaning that the selection of resistant mosquitoes will be slower than before."

Despite the worldwide ban, agricultural use continued in India, North Korea, and possibly elsewhere as of 2008.

Today, about 3,000 to 4,000 tons of DDT are produced each year for disease vector control. DDT is applied to the inside walls of homes to kill or repel mosquitoes. This intervention, called indoor residual spraying (IRS), greatly reduces environmental damage. It also reduces the incidence of DDT resistance. For comparison, treating 40 hectares (99 acres) of cotton during a typical U.S. growing season requires the same amount of chemical as roughly 1,700 homes.

Environmental Impact

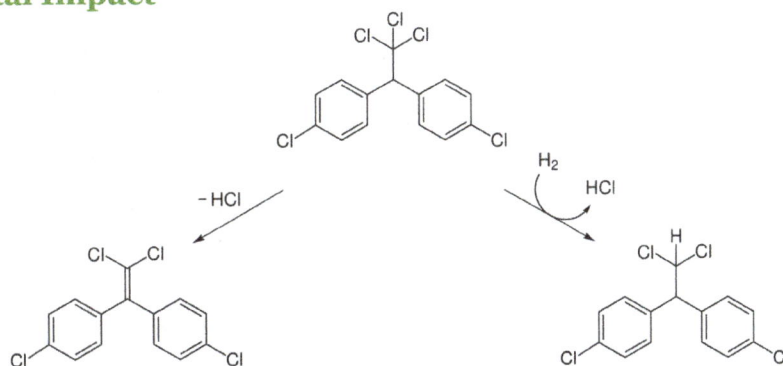

Degradation of DDT to form DDE (by elimination of HCl, left) and DDD (by reductive dechlorination, right)

DDT is a persistent organic pollutant that is readily adsorbed to soils and sediments, which can act both as sinks and as long-term sources of exposure affecting organisms. Depending on conditions, its soil half life can range from 22 days to 30 years. Routes of loss and degradation include runoff, volatilization, photolysis and aerobic and anaerobic biodegradation. Due to hydrophobic properties, in aquatic ecosystems DDT and its metabolites are absorbed by aquatic organisms and adsorbed on suspended particles, leaving little DDT dissolved in the water. Its breakdown products and metabolites, DDE and DDD, are also persistent and have similar chemical and physical properties. DDT and its breakdown products are transported from warmer areas to the Arctic by the phenomenon of global distillation, where they then accumulate in the region's food web.

Because of its lipophilic properties, DDT can bioaccumulate, especially in predatory birds. DDT, DDE and DDD magnify through the food chain, with apex predators such as raptor birds concentrating more chemicals than other animals in the same environment. They are stored mainly in body fat. DDT and DDE are resistant to metabolism; in humans, their half-lives are 6 and up to 10 years, respectively.

In the United States, these chemicals were detected in almost all human blood samples tested by the Centers for Disease Control in 2005, though their levels have sharply declined since most uses were banned. Estimated dietary intake has declined, although FDA food tests commonly detect it.

Marine macroalgae (seaweed) help reduce soil toxicity by up to 80% within six weeks.

Effects on Wildlife and Eggshell Thinning

DDT is toxic to a wide range of living organisms, including marine animals such as crayfish, daphnids, sea shrimp and many species of fish. DDE caused eggshell thinning and population declines in multiple North American and European bird of prey species. Eggshell thinning lowers the reproductive success rate of certain bird species by causing egg breakage and embryo deaths. DDE-related eggshell thinning is considered a major reason for the decline of the bald eagle, brown pelican, peregrine falcon and osprey. However, birds vary in their sensitivity to these chemicals. Birds of prey, waterfowl and song birds are more susceptible than chickens and related species. DDE appears to be more potent than DDT. Even in 2010, California condors that feed on sea lions at Big Sur that in turn feed in the Palos Verdes Shelf area of the Montrose Chemical Superfund site exhibited continued thin-shell problems. Scientists with the Ventana Wildlife Society and others study and remediate the condors' problems.

The biological thinning mechanism is not entirely understood, but strong evidence indictates that p,p'-DDE inhibits calcium ATPase in the membrane of the shell gland and reduces the transport of calcium carbonate from blood into the eggshell gland. This results in a dose-dependent thickness reduction. Other evidence indicates that o,p'-DDT disrupts female reproductive tract development, later impairing eggshell quality. Multiple mechanisms may be at work, or different mechanisms may operate in different species. Some studies show that although DDE levels have fallen dramatically, eggshell thickness remains 10–12 percent thinner than before DDT was first used.

Human Health

A U.S. soldier is demonstrating DDT hand-spraying equipment.
DDT was used to control the spread of typhus-carrying lice.

DDT is an endocrine disruptor. It is considered likely to be a human carcinogen although the majority of studies suggest it is not directly genotoxic. DDE acts as a weak androgen receptor antagonist, but not as an estrogen. p,p'-DDT, DDT's main component, has little or no androgenic or estrogenic activity. The minor component o,p'-DDT has weak estrogenic activity.

Acute Toxicity

DDT is classified as "moderately toxic" by the US National Toxicology Program (NTP) and "moderately hazardous" by WHO, based on the rat oral LD_{50} of 113 mg/kg. DDT has on rare occasions been administered orally as a treatment for barbiturate poisoning.

Chronic Toxicity

DDT and DDE, like other organochlorines, have been shown to have xenoestrogenic activity, meaning they are chemically similar enough to estrogens to trigger hormonal responses in animals. This endocrine disrupting activity has been observed in mice and rat toxicological studies. Epidemiological evidence indicates that these effects may be occurring in humans as a result of DDT exposure. EPA states that DDT exposure damages the reproductive system and reduces reproductive success. These effects may cause developmental and reproductive toxicity:

- A review article in *The Lancet* states, "research has shown that exposure to DDT at amounts that would be needed in malaria control might cause preterm birth and early weaning ... toxicological evidence shows endocrine-disrupting properties; human data also indicate possible disruption in semen quality, menstruation, gestational length, and duration of lactation."

- Other studies document decreases in semen quality among men with high exposures (generally from IRS).

- Studies generally find that high blood DDT or DDE levels do not increase time to pregnancy (TTP.) Some evidence indicates that the daughters of highly exposed women may have more increased TTP.

- DDT is associated with early pregnancy loss, a type of miscarriage. A prospective cohort study of Chinese textile workers found "a positive, monotonic, exposure-response association between preconception serum total DDT and the risk of subsequent early pregnancy losses." The median serum DDE level of study group was lower than that typically observed in women living in homes sprayed with DDT.

- A Japanese study of congenital hypothyroidism concluded that *in utero* DDT exposure may affect thyroid hormone levels and "play an important role in the incidence and/or causation of cretinism." Other studies found that DDT or DDE interfere with proper thyroid function in pregnancy and childhood.

- Exposure to DDT can cause shorter menstrual cycles.

Carcinogenicity

In 2002, the Centers for Disease Control and Prevention reported, "Overall, in spite of some positive associations for some cancers within certain subgroups of people, there is no clear evidence that exposure to DDT/DDE causes cancer in humans." The NTP classifies it as "reasonably anticipated to be a carcinogen," the International Agency for Research on Cancer classifies it as "probably carcinogenic to humans", and the EPA classifies DDT, DDE and DDD as class B2 "probable" carcinogens. These evaluations are based mainly on animal studies.

A 2005 Lancet review stated that occupational DDT exposure was associated with increased pancreatic cancer risk in 2 case control studies, but another study showed no DDE dose-effect association. Results regarding a possible association with liver cancer and biliary tract cancer are conflicting: workers who did not have direct occupational DDT contact showed increased risk. White men had an increased risk, but not white women or black men. Results about an association with multiple myeloma, prostate and testicular cancer, endometrial cancer and colorectal cancer have been inconclusive or generally do not support an association.

A 2009 review, whose co-authors included persons engaged in DDT-related litigation, reached broadly similar conclusions, with an equivocal association with testicular cancer. Case–control studies did not support an association with leukemia or lymphoma.

Breast Cancer

The question of whether DDT or DDE are risk factors in breast cancer has not been conclusively answered. Several meta analyses of observational studies have concluded that there is no overall relationship between DDT exposure and breast cancer risk. The United States Institute of Medicine reviewed data on the association of breast cancer with DDT exposure in 2012 and concluded that a causative relationship could neither be proven nor disproven.

A 2007 case control study using archived blood samples found that breast cancer risk was increased 5-fold among women who were born prior to 1931 and who had high serum DDT levels in 1963. Reasoning that DDT use became widespread in 1945 and peaked around 1950, they concluded that the ages of 14–20 were a critical period in which DDT exposure leads to increased risk. This study, which suggests a connection between DDT exposure and breast cancer that would not be picked up by most studies, has received variable commentary in third party reviews. One review suggested that "previous studies that measured exposure in older women may have missed the critical period." A second review suggested a cautious approach to the interpretation of these results given methodological weaknesses in the study design. The National Toxicology Program notes that while the majority of studies have not found a relationship between DDT exposure and breast cancer that positive associations have been seen in a "few studies among women with higher levels of exposure and among certain subgroups of women"

A 2015 case control study identified a link (odds ratio 3.4) between *in-utero* exposure (as estimated from archived maternal blood samples) and breast cancer diagnosis in daughters. The findings "support classification of DDT as an endocrine disruptor, a predictor of breast cancer, and a marker of high risk".

Malaria

Malaria remains the primary public health challenge in many countries. 2008 WHO estimates were 243 million cases and 863,000 deaths. About 89% of these deaths occur in Africa, mostly to children under age 5. DDT is one of many tools to fight the disease. Its use in this context has been called everything from a "miracle weapon [that is] like Kryptonite to the mosquitoes," to "toxic colonialism".

Before DDT, eliminating mosquito breeding grounds by drainage or poisoning with Paris green or pyrethrum was sometimes successful. In parts of the world with rising living standards, the elimination of malaria was often a collateral benefit of the introduction of window screens and improved sanitation. A variety of usually simultaneous interventions represents best practice. These include antimalarial drugs to prevent or treat infection; improvements in public health infrastructure to diagnose, sequester and treat infected individuals; bednets and other methods intended to keep mosquitoes from biting humans; and vector control strategies such as larvaciding with insecticides, ecological controls such as draining mosquito breeding grounds or introducing fish to eat larvae and indoor residual spraying (IRS) with insecticides, possibly including DDT. IRS involves the treatment of interior walls and ceilings with insecticides. It is particularly effective against mosquitoes, since many species rest on an indoor wall before or after feeding. DDT is one of 12 WHO–approved IRS insecticides.

WHO's anti-malaria campaign of the 1950s and 1960s relied heavily on DDT and the results were promising, though temporary in developing countries. Experts tie malarial resurgence to multiple factors, including poor leadership, management and funding of malaria control programs; poverty; civil unrest; and increased irrigation. The evolution of resistance to first-generation drugs (e.g. chloroquine) and to insecticides exacerbated the situation. Resistance was largely fueled by unrestricted agricultural use. Resistance and the harm both to humans and the environment led many governments to curtail DDT use in vector control and agriculture. In 2006 WHO reversed a longstanding policy against DDT by recommending that it be used as an indoor pesticide in regions where malaria is a major problem.

Once the mainstay of anti-malaria campaigns, as of 2008 only 12 countries used DDT, including India and some southern African states, though the number was expected to rise.

Initial Effectiveness

When it was introduced in World War II, DDT was effective in reducing malaria morbidity and mortality. WHO's anti-malaria campaign, which consisted mostly of spraying DDT and rapid treatment and diagnosis to break the transmission cycle, was initially successful as well. For example, in Sri Lanka, the program reduced cases from about one million per year before spraying to just 18 in 1963 and 29 in 1964. Thereafter the program was halted to save money and malaria rebounded to 600,000 cases in 1968 and the first quarter of 1969. The country resumed DDT vector control but the mosquitoes had evolved resistance in the interim, presumably because of continued agricultural use. The program switched to malathion, but despite initial successes, malaria continued its resurgence into the 1980s.

DDT remains on WHO's list of insecticides recommended for IRS. After the appointment of Arata Kochi as head of its anti-malaria division, WHO's policy shifted from recommending IRS only in areas of seasonal or episodic transmission of malaria, to advocating it in areas of continuous, intense transmission. WHO reaffirmed its commitment to phasing out DDT, aiming "to achieve a 30% cut in the application of DDT world-wide by 2014 and its total phase-out by the early 2020s if not sooner" while simultaneously combating malaria. WHO plans to implement alternatives to DDT to achieve this goal.

South Africa continues to use DDT under WHO guidelines. In 1996, the country switched to alternative insecticides and malaria incidence increased dramatically. Returning to DDT and introducing new drugs brought malaria back under control. Malaria cases increased in South America after countries in that continent stopped using DDT. Research data showed a strong negative relationship between DDT residual house sprayings and malaria. In a research from 1993 to 1995, Ecuador increased its use of DDT and achieved a 61% reduction in malaria rates, while each of the other countries that gradually decreased its DDT use had large increases.

Mosquito Resistance

In some areas resistance reduced DDT's effectiveness. WHO guidelines require that absence of resistance must be confirmed before using the chemical. Resistance is largely due to agricultural use, in much greater quantities than required for disease prevention.

Resistance was noted early in spray campaigns. Paul Russell, former head of the Allied Anti-Malaria campaign, observed in 1956 that "resistance has appeared after six or seven years." Resistance has been detected in Sri Lanka, Pakistan, Turkey and Central America and it has largely been replaced by organophosphate or carbamate insecticides, *e.g.* malathion or bendiocarb.

In many parts of India, DDT is ineffective. Agricultural uses were banned in 1989 and its anti-malarial use has been declining. Urban use ended. DDT is still manufactured and used. One study concluded that "DDT is still a viable insecticide in indoor residual spraying owing to its effectivity in well supervised spray operation and high excito-repellency factor."

Studies of malaria-vector mosquitoes in KwaZulu-Natal Province, South Africa found susceptibility to 4% DDT (WHO's susceptibility standard), in 63% of the samples, compared to the average of 86.5% in the same species caught in the open. The authors concluded that "Finding DDT resistance in the vector *An. arabiensis,* close to the area where we previously reported pyrethroid-resistance in the vector *An. funestus* Giles, indicates an urgent need to develop a strategy of insecticide resistance management for the malaria control programmes of southern Africa."

DDT can still be effective against resistant mosquitoes and the avoidance of DDT-sprayed walls by mosquitoes is an additional benefit of the chemical. For example, a 2007 study reported that resistant mosquitoes avoided treated huts. The researchers argued that DDT was the best pesticide for use in IRS (even though it did not afford the most protection from mosquitoes out of the three test chemicals) because the others pesticides worked primarily by killing or irritating mosquitoes – encouraging the development of resistance. Others argue that the avoidance behavior slows eradication. Unlike other insecticides such as pyrethroids, DDT requires long exposure to accumulate a lethal dose; however its irritant property shortens contact periods. "For these reasons, when comparisons have been made, better malaria control has generally been achieved with pyrethroids than with DDT." In India outdoor sleeping and night duties are common, implying that "the excito-repellent effect of DDT, often reported useful in other countries, actually promotes outdoor transmission." Genomic studies in the model genetic organism *Drosophila melanogaster* revealed that high level DDT resistance is polygenic, involving multiple resistance mechanisms.

Residents' Concerns

IRS is effective if at least 80% of homes and barns in a residential area are sprayed. Lower coverage rates can jeopardize program effectiveness. Many residents resist DDT spraying, objecting to the lingering smell, stains on walls, and the potential exacerbation of problems with other insect pests. Pyrethroid insecticides (e.g. deltamethrin and lambda-cyhalothrin) can overcome some of these issues, increasing participation.

Human Exposure

A 1994 study found that South Africans living in sprayed homes have levels that are several orders of magnitude greater than others. Breast milk from South African mothers contains high levels of DDT and DDE. It is unclear to what extent these levels arise from home spraying vs food residues. Evidence indicates that these levels are associated with infant neurological abnormalities.

Most studies of DDT's human health effects have been conducted in developed countries where DDT is not used and exposure is relatively low.

Illegal diversion to agriculture is also a concern as it is difficult to prevent and its subsequent use on crops is uncontrolled. For example, DDT use is widespread in Indian agriculture, particularly mango production and is reportedly used by librarians to protect books. Other examples include Ethiopia, where DDT intended for malaria control is reportedly used in coffee production, and Ghana where it is used for fishing." The residues in crops at levels unacceptable for export have been an important factor in bans in several tropical countries. Adding to this problem is a lack of skilled personnel and management.

Criticism of Restrictions on DDT Use

Critics argue that limitations on DDT use for public health purposes have caused unnecessary morbidity and mortality from vector-borne diseases, with some claims of malaria deaths ranging as high as the hundreds of thousands and millions. Robert Gwadz of the US National Institutes of Health said in 2007, "The ban on DDT may have killed 20 million children." These arguments were rejected as "outrageous" by former WHO scientist Socrates Litsios. May Berenbaum, University of Illinois entomologist, says, "to blame environmentalists who oppose DDT for more deaths than Hitler is worse than irresponsible." Investigative journalist Adam Sarvana and others characterize this notion as a "myth" promoted principally by Roger Bate of the pro-DDT advocacy group Africa Fighting Malaria (AFM).

Criticisms of a DDT "ban" often specifically reference the 1972 United States ban (with the erroneous implication that this constituted a worldwide ban and prohibited use of DDT in vector control). Reference is often made to *Silent Spring,* even though Carson never pushed for a DDT ban. John Quiggin and Tim Lambert wrote, "the most striking feature of the claim against Carson is the ease with which it can be refuted."

It has been alleged that donor governments and agencies refused to fund DDT spraying, or made aid contingent upon not using DDT. According to a report in the *British Medical Journal*, use of DDT in Mozambique "was stopped several decades ago, because 80% of the country's health budget came from donor funds, and donors refused to allow the use of DDT." Roger Bate asserted, "many countries have been coming under pressure from international health and environment agencies to give up DDT or face losing aid grants: Belize and Bolivia are on record admitting they gave in to pressure on this issue from [USAID]."

The US Agency for International Development (USAID) has been the focus of much criticism. While the agency now funds DDT use in some African countries, in the past it did not. When John Stossel accused USAID of not funding DDT because it wasn't "politically correct," Anne Peterson, the agency's assistant administrator for global health, replied that "I believe that the strategies we are using are as effective as spraying with DDT ... So, politically correct or not, I am very confident that what we are doing is the right strategy." USAID's Kent R. Hill stated that the agency had been misrepresented: "USAID strongly supports spraying as a preventative measure for malaria and will support the use of DDT when it is scientifically sound and warranted." The Agency's website states that "USAID has never had a 'policy' as such either 'for' or 'against' DDT for IRS (Indoor residual spraying). The real change in the past two years [2006/07] was a new interest and emphasis on IRS in general – with DDT or any other insecticide – as an effective malaria prevention strategy in tropical Africa." The agency claimed that in many cases alternative malaria control measures were more cost-effective than DDT spraying.

Alternatives

Insecticides

Organophosphate and carbamate insecticides, *e.g.* malathion and bendiocarb, respectively, are more expensive than DDT per kilogram and are applied at roughly the same dosage. Pyrethroids such as deltamethrin are also more expensive than DDT, but are applied more sparingly (0.02–0.3 g/m^2 vs 1–2 g/m^2), so the net cost per house is about the same.

Non-chemical Vector Control

Before DDT, malaria was successfully eliminated or curtailed in several tropical areas by removing or poisoning mosquito breeding grounds and larva habitats, for example by eliminating standing water. These methods have seen little application in Africa for more than half a century. According to CDC, such methods are not practical in Africa because "*Anopheles gambiae*, one of the primary vectors of malaria in Africa, breeds in numerous small pools of water that form due to rainfall ... It is difficult, if not impossible, to predict when and where the breeding sites will form, and to find and treat them before the adults emerge."

The relative effectiveness of IRS versus other malaria control techniques (e.g. bednets or prompt access to anti-malarial drugs) varies and is dependent on local conditions.

A WHO study released in January 2008 found that mass distribution of insecticide-treated mosquito nets and artemisinin–based drugs cut malaria deaths in half in malaria-burdened Rwanda and Ethiopia. IRS with DDT did not play an important role in mortality reduction in these countries.

Vietnam has enjoyed declining malaria cases and a 97% mortality reduction after switching in 1991 from a poorly funded DDT-based campaign to a program based on prompt treatment, bednets and pyrethroid group insecticides.

In Mexico, effective and affordable chemical and non-chemical strategies were so successful that the Mexican DDT manufacturing plant ceased production due to lack of demand.

A review of fourteen studies in sub-Saharan Africa, covering insecticide-treated nets, residual spraying, chemoprophylaxis for children, chemoprophylaxis or intermittent treatment for pregnant women, a hypothetical vaccine and changing front–line drug treatment, found decision making limited by the lack of information on the costs and effects of many interventions, the small number of cost-effectiveness analyses, the lack of evidence on the costs and effects of packages of measures and the problems in generalizing or comparing studies that relate to specific settings and use different methodologies and outcome measures. The two cost-effectiveness estimates of DDT residual spraying examined were not found to provide an accurate estimate of the cost-effectiveness of DDT spraying; the resulting estimates may not be good predictors of cost-effectiveness in current programs.

However, a study in Thailand found the cost per malaria case prevented of DDT spraying (US$1.87) to be 21% greater than the cost per case prevented of lambda-cyhalothrin–treated nets (US$1.54), casting some doubt on the assumption that DDT was the most cost-effective measure. The director of Mexico's malaria control program found similar results, declaring that it was 25% cheaper for Mexico to spray a house with synthetic pyrethroids than with DDT. However, another study in South Africa found generally lower costs for DDT spraying than for impregnated nets.

A more comprehensive approach to measuring cost-effectiveness or efficacy of malarial control would not only measure the cost in dollars, as well as the number of people saved, but would also consider ecological damage and negative human health impacts. One preliminary study found that it is likely that the detriment to human health approaches or exceeds the beneficial reductions in malarial cases, except perhaps in epidemics. It is similar to the earlier study regarding estimated theoretical infant mortality caused by DDT and subject to the criticism also mentioned earlier.

A study in the Solomon Islands found that "although impregnated bed nets cannot entirely replace DDT spraying without substantial increase in incidence, their use permits reduced DDT spraying."

A comparison of four successful programs against malaria in Brazil, India, Eritrea and Vietnam does not endorse any single strategy but instead states, "Common success factors included conducive country conditions, a targeted technical approach using a package of effective tools, data-driven decision-making, active leadership at all levels of government, involvement of communities, decentralized implementation and control of finances, skilled technical and managerial capacity at national and sub-national levels, hands-on technical and programmatic support from partner agencies, and sufficient and flexible financing."

DDT resistant mosquitoes have generally proved susceptible to pyrethroids. Thus far, pyrethroid resistance in *Anopheles* has not been a major problem.

Environmental Impact of Shipping

The environmental impact of shipping includes greenhouse gas emissions, acoustic, and oil pollution. The International Maritime Organization (IMO) estimates that Carbon dioxide emissions from shipping were equal to 2.2% of the global human-made emissions in 2012 and expects them to rise by as much as 2 to 3 times by 2050 if no action is taken.

A cargo ship discharging ballast water into the sea.

The First Intersessional Meeting of the IMO Working Group on Greenhouse Gas Emissions from Ships took place in Oslo, Norway on 23–27 June 2008. It was tasked with developing the technical basis for the reduction mechanisms that may form part of a future IMO regime to control greenhouse gas emissions from international shipping, and a draft of the actual reduction mechanisms themselves, for further consideration by IMO's Marine Environment Protection Committee (MEPC).

Ballast Water

Ballast water discharges by ships can have a negative impact on the marine environment.

Cruise ships, large tankers, and bulk cargo carriers use a huge amount of ballast water, which is often taken on in the coastal waters in one region after ships discharge wastewater or unload cargo, and discharged at the next port of call, wherever more cargo is loaded. Ballast water discharge typically contains a variety of biological materials, including plants, animals, viruses, and bacteria. These materials often include non-native, nuisance, invasive, exotic species that can cause extensive ecological and economic damage to aquatic ecosystems along with serious human health problems.

Sound Pollution

Noise pollution caused by shipping and other human enterprises has increased in recent history. The noise produced by ships can travel long distances, and marine species who may rely on sound for their orientation, communication, and feeding, can be harmed by this sound pollution

The Convention on the Conservation of Migratory Species has identified ocean noise as a potential threat to marine life. The disruption of whales' ability to communicate with one another is an extreme threat and is affecting their ability to survive. According to Discovery Channel's article on

Sonic Sea Journeys Deep Into the Ocean, over the last century, extremely loud noise from commercial ships, oil and gas exploration, naval sonar exercises and other sources has transformed the ocean's delicate acoustic habitat, challenging the ability of whales and other marine life to prosper and ultimately to survive. Whales are starting to react to this in ways that are life-threatening. Kenneth C. Balcomb, a whale researcher and a former U.S Navy officer, states that the day March 15, 2000, is the day of infamy. As Discovery says, where him and his crew discovered whales swimming dangerously close to the shore. They're supposed to be in deep water. So I pushed it back out to sea, says Balcomb. Although sonar helps to protect us, it is destroying marine life. According to IFAW Animal Rescue Program Director Katie Moore, There's different ways that sounds can affect animals. There's that underlying ambient noise level that's rising, and rising, and rising that interferes with communication and their movement patterns. And then there's the more acute kind of traumatic impact of sound, that's causing physical damage or a really strong behavioral response. It's fight or flight.

Wildlife Collisions

Marine mammals, such as whales and manatees, risk being struck by ships, causing injury and death. For example, if a ship is traveling at a speed of only 15 knots, there is a 79 percent chance of a collision being lethal to a whale.

One notable example of the impact of ship collisions is the endangered North Atlantic right whale, of which 400 or less remain. The greatest danger to the North Atlantic right whale is injury sustained from ship strikes. Between 1970 and 1999, 35.5 percent of recorded deaths were attributed to collisions. During 1999 to 2003, incidents of mortality and serious injury attributed to ship strikes averaged one per year. In 2004 to 2006, that number increased to 2.6. Deaths from collisions has become an extinction threat.

Atmospheric Pollution

Exhaust gases from ships are considered to be a significant source of air pollution, both for conventional pollutants and greenhouse gases.

There is a perception that cargo transport by ship is low in air pollutants, because for equal weight and distance it is the most efficient transport method, according to shipping researcher Amy Bows-Larkin. This is particularly true in comparison to air freight; however, because sea shipment accounts for far more annual tonnage and the distances are often large, shipping's emissions are globally substantial. A difficulty is that the year-on-year increasing amount shipping overwhelms gains in efficiency, such as from slow-steaming or the use of kites. The growth in tonne-kilometers of sea shipment has averaged 4 percent yearly since the 1990s. And it has grown by a factor of 5 since the 1970s. There are now over 100,000 transport ships at sea, of which about 6,000 are large container ships.

Conventional Pollutants

Air pollution from cruise ships is generated by diesel engines that burn high sulfur content fuel oil, also known as bunker oil, producing sulfur dioxide, nitrogen oxide and particulate, in addition to carbon monoxide, carbon dioxide, and hydrocarbons. Diesel exhaust has been classified by EPA as a likely human carcinogen. EPA recognizes that these emissions from marine diesel engines contribute to ozone and carbon monoxide non-attainment (i.e., failure to meet air quality standards), as well as adverse health effects associated with ambient concentrations of particulate matter and visibility, haze, acid deposition, and eutrophication and nitrification of water. EPA estimates that large marine diesel engines accounted for about 1.6 percent of mobile source nitrogen oxide emissions and 2.8 percent of mobile source particulate emissions in the United States in 2000. Contributions of marine diesel engines can be higher on a port-specific basis. Ultra-low sulfur diesel (ULSD) is a standard for defining diesel fuel with substantially lowered sulfur contents. As of 2006, almost all of the petroleum-based diesel fuel available in Europe and North America is of a ULSD type.

Of total global air emissions, shipping accounts for 18 to 30 percent of the nitrogen oxide and 9 percent of the sulphur oxides. Sulfur in the air creates acid rain which damages crops and buildings. When inhaled, sulfur is known to cause respiratory problems and even increases the risk of a heart attack. According to Irene Blooming, a spokeswoman for the European environmental coalition Seas at Risk, the fuel used in oil tankers and container ships is high in sulfur and cheaper to buy compared to the fuel used for domestic land use. "A ship lets out around 50 times more sulfur than a lorry per metric tonne of cargo carried." Cities in the U.S. like Long Beach, Los Angeles, Houston, Galveston, and Pittsburgh see some of the heaviest shipping traffic in the nation and have left local officials desperately trying to clean up the air. Increasing trade between the U.S. and China is helping to increase the number of vessels navigating the Pacific and exacerbating many of the environmental problems. To maintain the level of growth China is experiencing, large amounts of grain are being shipped to China by the boat load. The number of voyages are expected to continue increasing.

Greenhouse Gas Pollutants

3.5 to 4 percent of all climate change emissions are caused by shipping, primarily carbon dioxide.

Cruise ship haze over Juneau, Alaska

As one way to reduce the impact of greenhouse gas emissions from shipping, vetting agency Right-Ship developed an online "Greenhouse Gas (GHG) Emissions Rating" as a systematic way for the industry to compare a ship's CO_2 emissions with peer vessels of a similar size and type. Based on the International Maritime Organisation's (IMO) Energy Efficiency Design Index (EEDI) that applies to ships built from 2013, RightShip's GHG Rating can also be applied to vessels built prior to 2013, allowing for effective vessel comparison across the world's fleet. The GHG Rating utilises an A to G scale, where A represents the most efficient ships. It measures the theoretical amount of carbon dioxide emitted per tonne nautical mile travelled, based on the design characteristics of the ship at time of build such as cargo carrying capacity, engine power and fuel consumption. Higher rated ships can deliver significantly lower CO_2 emissions across the voyage length, which means they also use less fuel and are cheaper to run.

Stress for Improvement

One source of environmental stresses on maritime vessels recently has come from states and localities, as they assess the contribution of commercial marine vessels to regional air quality problems when ships are docked at port. For instance, large marine diesel engines are believed to contribute 7 percent of mobile source nitrogen oxide emissions in Baton Rouge/New Orleans. Ships can also have a significant impact in areas without large commercial ports: they contribute about 37 percent of total area nitrogen oxide emissions in the Santa Barbara area, and that percentage is expected to increase to 61 percent by 2015. Again, there is little cruise-industry specific data on this issue. They comprise only a small fraction of the world shipping fleet, but cruise ship emissions may exert significant impacts on a local scale in specific coastal areas that are visited repeatedly. Shipboard incinerators also burn large volumes of garbage, plastics, and other waste, producing ash that must be disposed of. Incinerators may release toxic emissions as well.

In 2005, MARPOL Annex VI came into force to combat this problem. As such cruise ships now employ CCTV monitoring on the smokestacks as well as recorded measuring via opacity meter while some are also using clean burning gas turbines for electrical loads and propulsion in sensitive areas.

Oil Spills

Most commonly associated with ship pollution are oil spills. While less frequent than the pollution that occurs from daily operations, oil spills have devastating effects. While being toxic to marine life, polycyclic aromatic hydrocarbons (PAHs), the components in crude oil, are very difficult to clean up, and last for years in the sediment and marine environment. Marine species constantly exposed to PAHs can exhibit developmental problems, susceptibility to disease, and abnormal reproductive cycles. One of the more widely known spills was the Exxon Valdez incident in Alaska. The ship ran aground and dumped a massive amount of oil into the ocean in March 1989. Despite efforts of scientists, managers and volunteers, over 400,000 seabirds, about 1,000 sea otters, and immense numbers of fish were killed.

International Regulation

Some of the major international efforts in the form of treaties are the Marine Pollution Treaty, Honolulu, which deals with regulating marine pollution from ships, and the UN Convention on Law of the Sea, which deals with marine species and pollution. While plenty of local and international regulations have been introduced throughout maritime history, much of the current regulations are considered inadequate. "In general, the treaties tend to emphasize the technical features of safety and pollution control measures without going to the root causes of sub-standard shipping, the absence of incentives for compliance and the lack of enforceability of measures." The most common problems encountered with international shipping arise from paperwork errors and customs brokers not having the proper information about your items. Cruise ships, for example, are exempt from regulation under the US discharge permit system (NPDES, under the Clean Water Act) that requires compliance with technology-based standards. In the Caribbean, many ports lack proper waste disposal facilities, and many ships dump their waste at sea.

Sewage

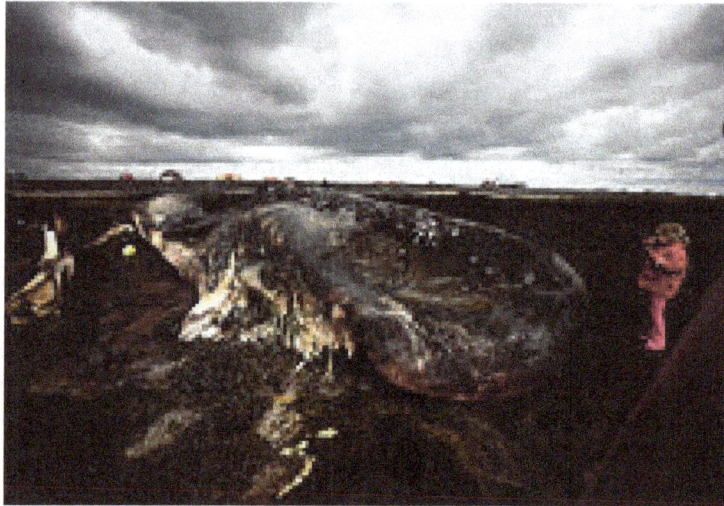

Carcass of a whale on a shore in Iceland.

The cruise line industry dumps 255,000 US gallons (970 m³) of greywater and 30,000 US gallons (110 m³) of blackwater into the sea every day. Blackwater is sewage, wastewater from toilets and medical facilities, which can contain harmful bacteria, pathogens, viruses, intestinal parasites, and harmful nutrients. Discharges of untreated or inadequately treated sewage can cause bacterial and viral contamination of fisheries and shellfish beds, producing risks to public health. Nutrients in sewage, such as nitrogen and phosphorus, promote excessive algal blooms, which consumes oxygen in the water and can lead to fish kills and destruction of other aquatic life. A large cruise ship (3,000 passengers and crew) generates an estimated 55,000 to 110,000 liters per day of blackwater waste.

Due to the environmental impact of shipping, and sewage in particular marpol annex IV was brought into force September 2003 strictly limiting untreated waste discharge. Modern cruise ships are most commonly installed with a membrane bioreactor type treatment plant for all blackwater and greywater, such as (http://www.gertsen-olufsen.com/Ship-Offshore/Products/G-O_Brands/G-O_Bioreactor.aspx) , Zenon or Rochem which produce near drinkable quality effluent to be re-used in the machinery spaces as technical water.

Cleaning

Greywater is wastewater from the sinks, showers, galleys, laundry, and cleaning activities aboard a ship. It can contain a variety of pollutant substances, including fecal coliforms, detergents, oil and grease, metals, organic compounds, petroleum hydrocarbons, nutrients, food waste, medical and dental waste. Sampling done by the EPA and the state of Alaska found that untreated greywater from cruise ships can contain pollutants at variable strengths and that it can contain levels of fecal coliform bacteria several times greater than is typically found in untreated domestic wastewater. Greywater has potential to cause adverse environmental effects because of concentrations of nutrients and other oxygen-demanding materials, in particular. Greywater is typically the largest source of liquid waste generated by cruise ships (90 to 95 percent of the total). Estimates of greywater range from 110 to 320 liters per day per person, or 330,000 to 960,000 liters per day for a 3,000-person cruise ship.

Solid Waste

Solid waste generated on a ship includes glass, paper, cardboard, aluminium and steel cans, and plastics. It can be either non-hazardous or hazardous in nature. Solid waste that enters the ocean may become marine debris, and can then pose a threat to marine organisms, humans, coastal communities, and industries that utilize marine waters. Cruise ships typically manage solid waste by a combination of source reduction, waste minimization, and recycling. However, as much as 75 percent of solid waste is incinerated on board, and the ash typically is discharged at sea, although some is landed ashore for disposal or recycling. Marine mammals, fish, sea turtles, and birds can be injured or killed from entanglement with plastics and other solid waste that may be released or disposed off of cruise ships. On average, each cruise ship passenger generates at least two pounds of non-hazardous solid waste per day. With large cruise ships carrying several thousand passengers, the amount of waste generated in a day can be massive. For a large cruise ship, about 8 tons of solid waste are generated during a one-week cruise. It has been estimated that 24 percent of the solid waste generated by vessels worldwide (by weight) comes from cruise ships. Most cruise ship garbage is treated on board (incinerated, pulped, or ground up) for discharge overboard. When garbage must be off-loaded (for example, because glass and aluminium cannot be incinerated), cruise ships can put a strain on port reception facilities, which are rarely adequate to the task of serving a large passenger vessel.

Bilge Water

On a ship, oil often leaks from engine and machinery spaces or from engine maintenance activities and mixes with water in the bilge, the lowest part of the hull of the ship, but there is a filter to clean bilge water before being discharged. Oil, gasoline, and by-products from the biological breakdown of petroleum products can harm fish and wildlife and pose threats to human health if ingested. Oil in even minute concentrations can kill fish or have various sub-lethal chronic effects. Bilge water also may contain solid wastes and pollutants containing high levels of oxygen-demanding material, oil and other chemicals. A typically large cruise ship will generate an average of 8 metric tons of oily bilge water for each 24 hours of operation. To maintain ship stability and eliminate potentially hazardous conditions from oil vapors in these areas, the bilge spaces need to be flushed and periodically pumped dry. However, before a bilge can be cleared out and the water discharged, the oil that has been accumulated needs to be extracted from the bilge water, after which the extracted oil can be reused, incinerated, and/or offloaded in port. If a separator, which is normally used to extract the oil, is faulty or is deliberately bypassed, untreated oily bilge water could be discharged directly into the ocean, where it can damage marine life. A number of cruise lines have been charged with environmental violations related to this issue in recent years.

Issues by Region

Asia

United Kingdom

- Merchant Shipping Act 1995

- Merchant Shipping (Pollution) Act 2006

United States

It is expected that, (from 2004) "...shipping traffic to and from the United States is projected to double by 2020."

- Act to Prevent Pollution from Ships

- American Bureau of Shipping

- Cruise ship pollution in the United States

- National Oil and Hazardous Substances Contingency Plan

- Oil Pollution Act of 1990

- Regulation of ship pollution in the United States

Committed Dose

The committed dose in radiological protection is a measure of the stochastic health risk due to an intake of radioactive material into the human body. Stochastic in this context is defined as the *probability* of cancer induction and genetic damage, due to low levels of radiation. The SI unit of measure is the sievert.

A committed dose from an internal source represents the same effective risk as the same amount of effective dose applied uniformly to the whole body from an external source, or the same amount of equivalent dose applied to part of the body. The committed dose is not intended as a measure for deterministic effects such as radiation sickness which is defined as the *severity* of a health effect which is certain to happen.

The radiation risk proposed by the International Commission on Radiological Protection (ICRP) predicts that an effective dose of one sievert carries a 5.5% chance of developing cancer. Such a risk is the sum of both internal and external radiation dose.

ICRP Definition

The ICRP states "Radionuclides incorporated in the human body irradiate the tissues over time periods determined by their physical half-life and their biological retention within the body. Thus they may give rise to doses to body tissues for many months or years after the intake. The need to regulate exposures to radionuclides and the accumulation of radiation dose over extended periods of time has led to the definition of committed dose quantities".

The ICRP defines two dose quantities for individual committed dose.

- Committed equivalent dose, H $_T$(t) is the time integral of the equivalent dose rate in a particular tissue or organ that will be received by an individual following intake of radioactive material into the body by a Reference Person, where t is the integration time in years. This refers specifically to the dose in a specific tissue or organ, in the similar way to external equivalent dose.

- Committed effective dose, E(t) is the sum of the products of the committed organ or tissue equivalent doses and the appropriate tissue weighting factors W_T, where t is the integration time in years following the intake. The commitment period is taken to be 50 years for adults, and to age 70 years for children. This refers specifically to the dose to the whole body, in the similar way to external effective dose. The committed effective dose is used to demonstrate compliance with dose limits and is entered into the "dose of record" for occupational exposures used for recording, reporting and retrospective demonstration of compliance with regulatory dose limits.

The ICRP further states "For internal exposure, committed effective doses are generally determined from an assessment of the intakes of radionuclides from bioassay measurements or other quantities (e.g., activity retained in the body or in daily excreta). The radiation dose is determined from the intake using recommended dose coefficients".

Dose Intake

The intake of radioactive material can occur through four pathways:

- inhalation of airborne contaminants such as radon

- ingestion of contaminated food or liquids

- absorption of vapours such as tritium oxide through the skin

- injection of medical radioisotopes such as technetium-99m

Some artificial radioisotopes such as iodine-131 are chemically identical to natural isotopes needed by the body, and may be more readily absorbed if the individual has a deficit of that element. For instance, Potassium iodide (KI), administered orally immediately after exposure, may be used to protect the thyroid from ingested radioactive iodine in the event of an accident or attack at a nuclear power plant, or the detonation of a nuclear explosive which would release radioactive iodine.

Other radioisotopes have an affinity for particular tissues, such as plutonium into bone, and may be retained there for years in spite of their foreign nature. In summary, not all radiation is harmful. The radiation can be absorbed through multiple pathways, varying due to the circumstances of the situation. If the radioactive material is necessary, it can be ingested orally via stable isotopes of specific elements. This is only suggested to those that have a lack of these elements however, because radioactive material can go from healthy to harmful with very small amounts. The most harmful way to absorb radiation is that of absorption because it is almost impossible to control how much will enter the body.

Physical Factors

Since irradiation increases with proximity to the source of radiation, and as it is impossible to distance or shield an internal source, radioactive materials inside the body can deliver much higher doses to the host organs than they normally would from outside the body. This is particularly true for alpha and beta emitters that are easily shielded by skin and clothing. Some have hypothesized that alpha's high relative biological effectiveness might be attributable to cell's tendency to absorb transuranic metals into the cellular nucleus where they would be in very close proximity to the genome, though an elevated effectiveness can also be observed for external alpha radiation in cellular studies. As in the calculations for equivalent dose and effective dose, committed dose must include corrections for the relative biological effectiveness of the radiation type and weightings for tissue sensitivity.

Duration

The dose rate from a single uptake decays over time due to both radioactive decay, and biological decay (i.e. excretion from the body). The combined radioactive and biological half-life, called the effective half-life of the material, may range from hours for medical radioisotopes to decades for transuranic waste. Committed dose is the integral of this decaying dose rate over the presumed remaining lifespan of the organism. Most regulations require this integral to be taken over 50 years for uptakes during adulthood or over 70 years for uptakes during childhood. In dosimetry accounting, the entire committed dose is conservatively assigned to the year of uptake, even though it may take many years for the tissues to actually accumulate this dose.

Measurement

There is no direct way to measure committed dose. Estimates can be made by analyzing the data from whole body counting, blood samples, urine samples, fecal samples, biopsies, and measurement of intake.

Whole body counting (WBC) is the most direct approach, but has some limitations: it cannot detect beta emitters such as tritium; it provides no chemical information about any compound that the radioisotope may be bound to; it may be inconclusive regarding the nature of the radioisotope detected; and it is a complex measurement subject to many sources of measurement and calibration error.

Analysis of blood samples, urine samples, fecal samples, and biopsies can provide more exact information about the chemical and isotopic nature of the contaminant, its distribution in the body, and the rate of elimination. Urine samples are the standard way to measure tritium intake, while fecal samples are the standard way to measure transuranic intake.

If the nature and quantity of radioactive materials taken into the body is known, and a reliable biochemical model of this material is available, this can be sufficient to determine committed dose. In occupational or accident scenarios, approximate estimates can be based on measurements of the environment that people were exposed to, but this cannot take into account factors such as breathing rate and adherence to hygiene practices. Exact information about the intake and its biochemical impact is usually only available in medical situations where radiopharmaceuticals are measured in a radioisotope dose calibrator prior to injection.

Annual limit on intake (ALI) is the derived limit for the amount of radioactive material taken into the body of an adult worker by inhalation or ingestion in a year. ALI is the intake of a given radio-nuclide in a year that would result in:

- a committed effective dose equivalent of 0.02 Sv (2 rems) for a "reference human body", or

- a committed dose equivalent of 0.2 Sv (20 rems) to any individual organ or tissue,

whatever dose is the smaller.

Health Effects

Intake of radioactive materials into the body tends to increase the risk of cancer, and possibly other stochastic effects. The International Commission on Radiological Protection has proposed a model whereby the incidence of cancers increases linearly with effective dose at a rate of 5.5% per sievert. This model is widely accepted for external radiation, but its application to internal contamination has been disputed. This model fails to account for the low rates of cancer in early workers at Los Alamos National Laboratory who were exposed to plutonium dust, and the high rates of thyroid cancer in children following the Chernobyl accident. The informal European Committee on Radiation Risk has questioned the ICRP model used for internal exposure.[unreliable source?] However a UK National Radiological Protection Board report endorses the ICRP approaches to the estimation of doses and risks from internal emitters and agrees with CERRIE conclusions that these should be best estimates and that associated uncertainties should receive more attention.

The true relationship between committed dose and cancer is almost certainly non-linear. For example, iodine-131 is notable in that high doses of the isotope are sometimes less dangerous than low doses, since they tend to kill thyroid tissues that would otherwise become cancerous as a result of the radiation. Most studies of very-high-dose I-131 for treatment of Graves disease have failed to find any increase in thyroid cancer, even though there is linear increase in thyroid cancer risk with I-131 absorption at moderate doses.

Internal exposure of the public is controlled by regulatory limits on the radioactive content of food and water. These limits are typically expressed in becquerel/kilogram, with different limits set for each contaminant.

Intake of very large amounts of radioactive material can cause acute radiation syndrome (ARS) in rare instances. Examples include the Alexander Litvinenko poisoning and Leide das Neves Ferreira. While there is no doubt that internal contamination was the cause of ARS in these cases, there is not enough data to establish what quantities of committed dose might cause ARS symptoms. In most scenarios where ARS is a concern, the external effective radiation dose is usually much more hazardous than the internal dose. Normally, the greatest concern with internal exposure is that the radioactive material may stay in the body for an extended period of time, "committing" the subject to accumulating dose long after the initial exposure has ceased. Over a hundred people, including

Eben Byers and the radium girls, have received committed doses in excess of 10 Gy and went on to die of cancer or natural causes, whereas the same amount of acute external dose would invariably cause an earlier death by ARS.

Examples

Below are a series of examples of internal exposure.

- Thorotrast

- The exposure caused by Potassium-40 present within a *normal* person.

- The exposure to the ingestion of a soluble radioactive substance, such as ^{89}Sr in cows' milk.

- A person who is being treated for cancer by means of an *unsealed source* radiotherapy method where a radioisotope is used as a drug (usually a liquid or pill). A review of this topic was published in 1999. Because the radioactive material becomes intimately mixed with the affected object it is often difficult to decontaminate the object or person in a case where internal exposure is occurring. While some very insoluble materials such as fission products within a uranium dioxide matrix might never be able to truly become part of an organism, it is normal to consider such particles in the lungs and digestive tract as a form of internal contamination which results in internal exposure.

- Boron neutron capture therapy (BNCT) involves injecting a boron-10 tagged chemical that preferentially binds to tumor cells. Neutrons from a nuclear reactor are shaped by a neutron moderator to the neutron energy spectrum suitable for BNCT treatment. The tumor is selectively bombarded with these neutrons. The neutrons quickly slow down in the body to become low energy *thermal neutrons*. These *thermal neutrons* are captured by the injected boron-10, forming excited (boron-11) which breaks down into lithium-7 and a helium-4 alpha particle both of these produce closely spaced ionizing radiation.This concept is described as a binary system using two separate components for the therapy of cancer. Each component in itself is relatively harmless to the cells, but when combined together for treatment they produce a highly cytocidal (cytotoxic) effect which is lethal (within a limited range of 5-9 micrometers or approximately one cell diameter). Clinical trials, with promising results, are currently carried out in Finland and Japan.

Related Quantities

The US Nuclear Regulatory commission defines some non-SI quantities for the calculation of committed dose for use only within the US regulatory system. They carry different names to those used within the International ICRP radiation protection system, thus:

- Committed dose equivalent (CDE) is the equivalent dose received by a particular organ or tissue from an internal source, without weighting for tissue sensitivity. This is essentially an intermediate calculation result that cannot be directly compared to final dosimetry quantities

- Committed effective dose equivalent (CEDE) as defined in Title 10, Section 20.1003, of the Code of Federal Regulations of the USA the CEDE dose (HE,50) is the sum of the products of the committed dose equivalents for each of the body organs or tissues that are irradiated multiplied by the weighting factors (WT) applicable to each of those organs or tissues.

Confusion between US and ICRP dose quantity systems can arise because the use of the term "dose equivalent" has been used within the ICRP system since 1991 only for quantities calculated using the value of Q (Linear energy transfer - LET), which the ICRP calls "operational quantities". However within the US NRC system "dose equivalent" is still used to name quantities which are calculated with tissue and radiation weighting factors, which in the ICRP system are now known as the "protection quantities" which are called "effective dose" and "equivalent dose".

Methylcyclopentadienyl Manganese Tricarbonyl

Methylcyclopentadienyl manganese tricarbonyl (MMT or MCMT) is an organomanganese compound with the formula $(CH_3C_5H_4)Mn(CO)_3$. Initially marketed as a supplement for use in leaded gasoline, MMT was later used in unleaded gasoline to increase the octane rating. Following the implementation of the Clean Air Act (United States) (CAA) in 1970, MMT continued to be used alongside tetraethyl lead (TEL) in the US as leaded gasoline was phased out (prior to TEL finally being banned from US gasoline in 1995), and was also used in unleaded gasoline until 1977. Ethyl Corporation obtained a waiver from the U.S. EPA (Environmental Protection Agency) in 1995, which allows the use of MMT in US unleaded gasoline (not including reformulated gasoline) at a treat rate equivalent to 8.3 mg Mn/L (manganese per liter).

MMT has been used in Canadian gasoline since 1976 (and in numerous other countries for many years) at a concentration up to 18 mg Mn/L (though the importation and interprovincial trade of gasoline containing MMT was restricted briefly during the period 1997-1998) and was introduced into Australia in 2000. It has been sold under the tradenames HiTEC® 3000, Cestoburn and Ecotane.

History of Use in the U.S.

Though initially marketed in 1958 as a smoke suppressant for gas turbines, MMT was further developed as an octane improver in 1974. When the United States Environmental Protection Agency (EPA) ordered the phase out of TEL in gasoline in 1973, new fuel additives were sought. TEL has been, and still is, used in certain countries as an additive to increase the octane rating of automotive gasoline.

In 1977, the US Congress amended the CAA to require advance approval by the EPA for the continued use of fuel additives such as MMT, ethanol, ethyl tert-butyl ether (ETBE), etc. The new CAA amendment required a "waiver" to allow use of fuel additives made of any elements other than carbon, hydrogen, oxygen (within certain limits) and nitrogen. To obtain a waiver, the applicant was required to demonstrate that the fuel additive would not lead to a failure of vehicle emission control systems.

Ethyl Corporation (Ethyl) applied to the US EPA for a waiver for MMT in both 1978 and 1981; in both cases the applications were denied because of stated concerns that MMT might damage catalytic converters and increase hydrocarbon emissions. In 1988, Ethyl began a new series of discussions with the EPA to determine a program for developing the necessary data to support a waiver application. In 1990, Ethyl filed its third waiver application prompting an extensive four-year review process. In 1993, the U.S. EPA determined that use of MMT at 8.3 mg Mn/l would not cause, or contribute to, vehicle emission control system failures.

Despite that finding, the EPA ultimately denied the waiver request in 1994 due to uncertainty related to health concerns regarding manganese emissions from the use of MMT.

As a result of this ruling, Ethyl initiated a legal action claiming that the EPA had exceeded its authority by denying the waiver on these grounds. This was upheld by the US Court of Appeals and EPA subsequently granted a waiver which allows the use of MMT in US unleaded gasoline (not including reformulated gasoline) at a treat rate equivalent to 8.3 mg Mn/l.

Implementation of this less-toxic alternative to TEL has been controversial. Opposition from automobile manufacturers and some areas of the scientific community has reportedly prompted oil companies to voluntarily stop the use of MMT in some of their countries of operation.

MMT is currently manufactured in the U.S. by the Afton Chemical Corporation, a subsidiary of New Market Corporation. It is also produced and marketed as Cestoburn, by Cestoil Chemical Inc. in Canada.

Structure and Synthesis

MMT is manufactured by reduction of bis(methylcyclopentadienyl) manganese using triethylaluminium. The reduction is conducted under an atmosphere of carbon monoxide. The reaction is exothermic, and without proper cooling, can lead to catastrophic thermal runaway.

MMT is a so-called half-sandwich compound, or more specifically a piano-stool complex (since the three CO ligands are like the legs of a piano stool). The manganese atom in MMT is coordinated with three carbonyl groups as well as to the methylcyclopentadienyl ring. These hydrophobic organic ligands make MMT highly lipophilic.

Related Compounds

A variety of related complexes are known, including ferrocene, which has also been used as an additive to gasoline.

Although of no practical value, the related compound cyclopentadienyl manganese tricarbonyl $(C_5H_5)Mn(CO)_3$ is also well studied. Up to two of the CO ligands in MMT can be replaced with thiocarbonyl groups, as illustrated by the compounds $(CH_3C_5H_4)Mn(CS)_2CO$ and $(CH_3C_5H_4)Mn(CS)(CO)_2$.

Safety

The human and environmental health impacts that may result from the use of MMT will be a function of exposure to either: (1) MMT in its original, unchanged, chemical form and/or (2) manganese combustion products emitted from vehicles operating on gasoline containing MMT as an octane improver.

MMT (As a Chemical Before Combustion in Gasoline)

The general public has minimal to zero direct exposure to MMT as a chemical, before it is combusted in gasoline. As stated by the US EPA in their risk assessment on MMT, "except for accidental or occupational contacts, exposure to MMT itself was not thought likely to pose a significant risk to the general population." Similarly, the Australian National Industrial Chemicals Notification and Assessment Scheme (NICNAS) stated that "[m]inimal public exposure to MMT is likely as a result of spills and splashes of LRP [lead replacement petrol] and aftermarket additives".

The MMT dossier registered in the European Chemical Agency's webpage indicates that before combustion in gasoline, MMT is classified as an acute toxicant by the oral, dermal, and inhalation routes of exposure under the European Union's Classification, Labeling and Packaging Regulation (EC/1272/2008), implementing the Global Harmonized System (GHS) of Classification and Labeling. The US ATSDR (Agency for Toxic Substances and Disease Registry) notes that MMT is very unstable in light and degrades to a mixture of less harmful substances and inorganic manganese in less than 2 minutes. Therefore, human exposure to MMT prior to combustion in gasoline would not likely occur at significant levels.

Regarding occupational exposure to the raw concentrated chemical prior to addition in gasoline, it has been noted that acute exposures to high levels of MMT in its raw concentrated form, prior to addition in gasoline, have resulted in giddiness, headache, nausea, chest tightness, dyspnea and paresthesia. In animals, acute lethal exposure to MMT is associated with damage to the lungs, kidney, liver and spleen, as well as tremors, convulsions, dyspnea and weakness. In both animals and humans, slight skin and eye irritation may result from dermal and ocular exposure, respectively. Data show that repeated inhalation exposure to MMT in rats results in histological changes in the lungs at levels greater than 3 mg/m^3. No effects were seen in the lungs or brain of monkeys when treated with up to 30 mg/m^3 MMT.

Chronic exposure to high levels of manganese, typically in certain occupational activities (such as welding), has also been known to cause manganism, a rare disease with symptoms similar to those of Parkinson's disease. Manganism has generally been eliminated due to proper controls in these occupational settings, such as in the ferro-alloy industry.

The US OSHA (Occupational Health and Safety Administration) has not established a permissible exposure limit specifically for MMT. However, OSHA has set a permissible exposure limit at a ceiling of 5 mg/m^3 for manganese and its compounds, while the National Institute for Occupational Safety and Health recommends workers not be exposed to more than 0.2 mg/m^3, over an eight-hour time-weighted average. In Europe, the MMT DNELs (Derived No Effect Level) for workers by the inhalation and dermal routes of exposure are 0.6 mg/m^3 and 0.11 mg/kg-day, respectively. The MMT DNELs for the general population by the inhalation and dermal routes of exposure are 0.11 mg/m^3 and 0.062 mg/kg-day, respectively.

In summary, based on the low potential for MMT release under normal storage and use, and MMT's rapid photo-degradation properties, environmental exposures are expected to be minimal. This is emphasized again by NICNAS in their conclusions which state that "[u]se of MMT in internal combustion engines as a fuel additive and subsequent degradation through combustion, and its short persistence in the environment, indicate that aquatic and terrestrial organisms are unlikely to be exposed to MMT at or above levels of concern through existing use as an AVSR. A low environmental risk is predicted".

Combustion Products

The health hazards associated with manganese compounds (manganese phosphate, manganese sulphate and manganese tetraoxide) emitted from vehicles operating on gasoline containing MMT have been debated for decades. In 1994 (reaffirmed in 1998, 2001 and 2010), Health Canada concluded that "airborne manganese resulting from the combustion of MMT in gasoline powered vehicles is not entering the Canadian environment in quantities or under conditions that may constitute a health risk" and confirmed they were taking no action with respect to MMT. Similarly, the 2003 NICNAS report ruled that the airborne concentrations of manganese as a result of car emissions from vehicles using fuel containing MMT poses no health hazard.

Additional health studies, overseen by the US EPA, were conducted to explain the transport of manganese in the body. The most recent of these studies were published by the Hamner Institutes for Health Sciences from 2007 through 2011 and submitted to the US EPA under the framework of the CAA . They show that manganese is naturally present in the environment and the body's natural mechanisms can handle a wide range of manganese intake via inhalation or ingestion, such that no significant health effects are anticipated from the use of MMT. While the body's natural mechanisms can be overwhelmed if exposures to manganese are very high, the testing confirms that the body can safely handle inhaled manganese at, and well above, levels observed when MMT is used in gasoline. The studies also indicate that inorganic manganese levels that result from use of MMT are safe for the entire population including vulnerable groups such as infants and the elderly. Although the EPA will now formally review the research findings, the US Agency recommended this data as meeting the Clean Air Act's health study's requirements and objectives.

Similar to the above description of human health risks, the environmental risks from MMT combustion products are also expected to be insignificant. In the assessment conducted by NICNAS, it was concluded that "[m]anganese, the principle degradation by-product from combustion of MMT, is naturally occurring and ubiquitous in the environment. It is an essential nutrient of plants and animals. Environmental exposure to Mn compounds will mostly arise through the gaseous phase. Eventually, these will deposit to land and waters. The emission of Mn into the environment from use of fuels containing MMT is unlikely to develop to levels of concern and therefore poses a low risk for terrestrial or aquatic environments."

Overall Combined Risk Assessment

Based on the low potential for the release of concentrated MMT (before it's combustion in gasoline) under normal storage and use, as well as its rapid photo-degradation properties, it has been concluded in multiple technical and global regulatory assessments that significant impacts to human health or the environment from MMT use are not anticipated. NICNAS concluded that there is "low occupational risk associated with MMT" both "for workers involved in formulating and distributing LRP or aftermarket fuel additives and those involved in automotive maintenance". Further, they also concluded that there is a "low risk" to the public from the use of MMT.

Significant human or environmental exposures associated with manganese compounds (manganese phosphate, manganese sulphate and manganese tetraoxide) from the combustion of MMT are not expected. In Health Canada's risk assessment on the health implications of the manganese combustion products of MMT, it was concluded that manganese exposures from MMT use are unlikely to pose a risk to health for any sub-group of the population. NICNAS similarly concluded that chronic Mn exposures (from all sources combined) are unlikely to be significantly changed by the use of MMT as a fuel additive.

In 2013, a risk assessment on MMT was developed by ARCADIS Consulting and verified by an independent panel, according to the methodology provided by the European Commission in compliance with the requirements of the European Fuel Quality Directive (2009/30/EC). The conclusions of this risk assessment are that "for MMT and its transformation products, when MMT is used as a fuel additive in petrol, no significant human health or environmental concerns related to exposure to either MMT or its transformation [combustion] products (manganese phosphate, manganese sulphate and manganese tetroxide) were identified at use at levels up to 18 mg Mn/l. Depending on the regional needs and the vehicle emission control technology available, an MMT treat rate in the range of 8.3 mg Mn/l to 18 mg Mn/l is scientifically justified and may deliver both environmental and economic benefits without significant adverse effects."

Based on the scientific work performed in the last 20 years on MMT and its combustion products, the use of MMT in fuel at levels up to 18 mg/l does not represent a significant risk to human health or the environment.

Vehicle Manufacturer Recommendations

Some manufacturers recommend against use of MMT in their vehicles, while others specifically prohibit its use.

Methylmercury

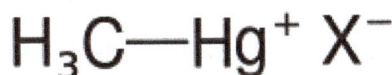

$$H_3C-Hg^+ \ X^-$$

Formula. X stands for any anion

3D model

Methylmercury (sometimes methyl mercury) is an organometallic cation with the formula [CH$_3$Hg]$^+$. It is a bioaccumulative environmental toxicant.

Structure

"Methylmercury" is a shorthand for "monomethylmercury", and is more correctly *"monomethyl-mercury(II) cation"*. It is composed of a methyl group (CH_3-) bonded to a mercury ion; its chemical formula is CH_{3Hg}^+ (sometimes written as $MeHg^+$). As a positively charged ion it readily combines with anions such as chloride (Cl^-), hydroxide (OH^-) and nitrate (NO_3^-). It also has very high affinity for sulfur-containing anions, particularly the thiol (-SH) groups on the amino acid cysteine and hence in proteins containing cysteine, forming a covalent bond. More than one cysteine moiety may coordinate with methylmercury, and methylmercury may migrate to other metal-binding sites in proteins.

Sources

Environmental Sources

Methylmercury is formed from inorganic mercury by the action of microbes that live in aquatic systems including lakes, rivers, wetlands, sediments, soils and the open ocean. Natural sources of mercury to the atmosphere include volcanoes, forest fires, volatization from the ocean and weathering of mercury-bearing rocks.

In the past, methylmercury was produced directly and indirectly as part of several industrial processes such as the manufacture of acetaldehyde. Currently there are few anthropogenic sources of methylmercury pollution in the United States other than as an indirect consequence of the burning of wastes containing inorganic mercury and from the burning of fossil fuels, particularly coal. Although inorganic mercury is only a trace constituent of such fuels, their large scale combustion in utility and commercial/industrial boilers in the United States alone results in release of some 80.2 tons (73 tonnes) of elemental mercury to the atmosphere each year, out of total anthropogenic mercury emissions in the United States of 158 tons (144 tonnes)/year.

Flooding of soils associated with reservoir creation (e.g. for hydroelectric power generation) has been linked to increased methylmercury concentrations in reservoir water and fish.

Methylmercury production in inland and marine ecosystems has been primarily attributed to anaerobic bacteria in the sediment. However, peaks in methylmercury in ocean water column and strong associations between methylmercury, nutrients and organic matter remineralization suggest water column production of methylmercury during carbon remineralization. Direct measurements of methylmercury production using stable mercury isotopes were made in oxic waters, but the microbes involved are still unknown.

Acute methylmercury poisoning occurred at Grassy Narrows in Ontario, Canada as a result of mercury released from the mercury-cell Chloralkali process, which uses liquid mercury as an electrode in a process that entails electrolytic decomposition of brine, followed by mercury methylation in the aquatic environment. An acute methylmercury poisoning tragedy occurred in Minamata, Japan following release of methylmercury into Minamata Bay and its tributaries. In the Ontario case, inorganic mercury discharged into the environment was methylated in the environment; whereas in Minamata, Japan, there was direct industrial discharge of methylmercury.

Dietary Sources

Because methylmercury is formed in aquatic systems and because it is not readily eliminated from organisms it is biomagnified in aquatic food chains from bacteria, to plankton, through macroinvertebrates, to herbivorous fish and to piscivorous (fish-eating) fish. At each step in the food chain, the concentration of methylmercury in the organism increases. The concentration of methylmercury in the top level aquatic predators can reach a level a million times higher than the level in the water. This is because methylmercury has a half-life of about 72 days in aquatic organisms resulting in its bioaccumulation within these food chains. Organisms, including humans, fish-eating birds, and fish-eating mammals such as otters and whales that consume fish from the top of the aquatic food chain receive the methylmercury that has accumulated through this process. Fish and other aquatic species are the only significant source of human methylmercury exposure.

The concentration of mercury in any given fish depends on the species of fish, the age and size of the fish and the type of water body in which it is found. In general, fish-eating fish such as shark, swordfish, marlin, larger species of tuna, walleye, largemouth bass, and northern pike, have higher levels of methylmercury than herbivorous fish or smaller fish such as tilapia and herring. Within a given species of fish, older and larger fish have higher levels of methylmercury than smaller fish. Fish that develop in water bodies that are more acidic also tend to have higher levels of methylmercury.

Biological Impact

Human Health Effects

Ingested methylmercury is readily and completely absorbed by the gastrointestinal tract. It is mostly found complexed with free cysteine and with proteins and peptides containing that amino acid. The methylmercuric-cysteinyl complex is recognized by amino acid transporting proteins in the body as methionine, another essential amino acid. Because of this mimicry, it is transported freely throughout the body including across the blood–brain barrier and across the placenta, where it is absorbed by the developing fetus. Also for this reason as well as its strong binding to proteins, methylmercury is not readily eliminated. Methylmercury has a half-life in human blood of about 50 days.

Several studies indicate that methylmercury is linked to subtle developmental deficits in children exposed in-utero such as loss of IQ points, and decreased performance in tests of language skills, memory function and attention deficits. Methylmercury exposure in adults has also been linked to increased risk of cardiovascular disease including heart attack. Some evidence also suggests that methylmercury can cause autoimmune effects in sensitive individuals. Despite some concerns about the relationship between methylmercury exposure and autism, there are few data that support such a link. Although there is no doubt that methylmercury is toxic in several respects, including through exposure of the developing fetus, there is still some controversy as to the levels of methylmercury in the diet that can result in adverse effects. Recent evidence suggests that the developmental and cardiovascular toxicity of methylmercury may be mitigated by co-exposures to omega-3 fatty acids and perhaps selenium, both found in fish and elsewhere

There have been several episodes in which large numbers of people were severely poisoned by food contaminated with high levels of methylmercury, notably the dumping of industrial waste that resulted in the pollution and subsequent mass poisoning in Minamata and Niigata, Japan and the situation in Iraq in the 1960s and 1970s in which wheat treated with methylmercury as a preservative and intended as seed grain was fed to animals and directly consumed by people. These episodes resulted in neurological symptoms including paresthesias, loss of physical coordination, difficulty in speech, narrowing of the visual field, hearing impairment, blindness, and death. Children who had been exposed in-utero through their mothers' ingestion were also affected with a range of symptoms including motor difficulties, sensory problems and intellectual disability.

At present, exposures of this magnitude are rarely seen and are confined to isolated incidents. Accordingly, concern over methylmercury pollution is currently focused on more subtle effects that may be linked to levels of exposure presently seen in populations with high to moderate levels of dietary fish consumption. These effects are not necessarily identifiable on an individual level or may not be uniquely recognizable as due to methylmercury. However, such effects may be detected by comparing populations with different levels of exposure. There are isolated reports of various clinical health effects in individuals who consume large amounts of fish; however, the specific health effects and exposure patterns have not been verified with larger, controlled studies.

Many governmental agencies, the most notable ones being the United States Environmental Protection Agency (EPA), the United States Food and Drug Administration (FDA), Health Canada, and the European Union Health and Consumer Protection Directorate-General, as well as the World Health Organization (WHO) and the United Nations Food and Agriculture Organization (FAO), have issued guidance for fish consumers that is designed to limit methylmercury exposure from fish consumption. At present, most of this guidance is based on protection of the developing fetus; future guidance, however, may also address cardiovascular risk. In general, fish consumption advice attempts to convey the message that fish is a good source of nutrition and has significant health benefits, but that consumers, in particular pregnant women, women of child-bearing age, nursing mothers, and young children, should avoid fish with high levels of methylmercury, limit their intake of fish with moderate levels of methylmercury, and consume fish with low levels of methylmercury no more than twice a week.

Effects on Fish and Wildlife

In recent years, there has been increasing recognition that methylmercury affects fish and wildlife health, both in acutely polluted ecosystems and ecosystems with modest methylmercury levels. Two reviews document numerous studies of diminished reproductive success of both fish, fish-eating birds, and mammals due to methylmercury contamination in aquatic ecosystems.

In Public Policy

Methylmercury contamination in fish, along with fish consumption advisories, has the potential to disrupt people's eating habits, fishing traditions, and the livelihoods of people involved in the capture, distribution, and preparation of fish as a foodstuff for humans. Furthermore, proposed limits on mercury emissions have the potential to add costly pollution controls on coal-fired utility boilers. Therefore, the methylmercury issue has attracted the attention of many levels of government, environmental groups, consumer advocates, science groups, food-industry-funded groups that question the science, and significant interest from electric utilities.

Carbon Monoxide Poisoning

Carbon monoxide poisoning occurs after too much inhalation of carbon monoxide (CO). Carbon monoxide is a toxic gas, but, being colorless, odorless, tasteless, and initially non-irritating, it is very difficult for people to detect. Carbon monoxide is a product of incomplete combustion of organic matter due to insufficient oxygen supply to enable complete oxidation to carbon dioxide (CO_2). It is often produced in domestic or industrial settings by motor vehicles that run on gasoline, diesel, methane, or other carbon-based fuels and from tools, gas heaters, and cooking equipment that are powered by carbon-based fuels such as propane, butane and charcoal. Exposure at 100 ppm or greater can be dangerous to human health.

Symptoms of mild acute poisoning include lightheadedness, confusion, headache, vertigo, and flu-like effects; larger exposures can lead to significant toxicity of the central nervous system and heart, and death. After acute poisoning, long-term sequelae often occur. Carbon monoxide can also have severe effects on the fetus of a pregnant woman. Chronic exposure to low levels of carbon monoxide can lead to depression, confusion, and memory loss. Carbon monoxide mainly causes adverse effects in humans by combining with hemoglobin to form carboxyhemoglobin (HbCO) in the blood. This prevents hemoglobin from carrying oxygen to the tissues, effectively reducing the oxygen-carrying capacity of the blood, leading to hypoxia. Additionally, myoglobin and mitochondrial cytochrome oxidase are thought to be adversely affected. Carboxyhemoglobin can revert to hemoglobin, but the recovery takes time because the HbCO complex is fairly stable.

Treatment of poisoning largely consists of administering 100% oxygen or providing hyperbaric oxygen therapy, although the optimum treatment remains controversial. Oxygen works as an antidote as it increases the removal of carbon monoxide from hemoglobin, in turn providing the body with normal levels of oxygen. The prevention of poisoning is a significant public health issue. Domestic carbon monoxide poisoning can be prevented by early detection with the use of household carbon monoxide detectors. Carbon monoxide poisoning is the most common type of fatal poisoning in many countries. Historically, it was also commonly used as a method to commit suicide, usually by deliberately inhaling the exhaust fumes of a running car engine. Modern automobiles, even with electronically controlled combustion and catalytic converters, can still produce levels of carbon monoxide which will kill if enclosed within a garage or if the tailpipe is obstructed (for example, by snow) and exhaust gas cannot escape normally. Carbon monoxide poisoning has also been speculated as a possible cause of apparent haunted houses; symptoms such as delirium and hallucinations may have led people suffering poisoning to think they have seen ghosts or to believe their house is haunted.

Signs and Symptoms

Carbon monoxide is not toxic to all forms of life. Its harmful effects are due to binding with hemoglobin so its danger to organisms that do not use this compound is doubtful. It thus has no effect on photosynthesising plants. It is easily absorbed through the lungs. Inhaling the gas can lead to hypoxic injury, nervous system damage, and even death. Different people and populations may have different carbon monoxide tolerance levels. On average, exposures at 100 ppm or greater is dangerous to human health. In the United States, the OSHA limits long-term workplace exposure levels to less than 50 ppm averaged over an 8-hour period; in addition, employees are to be removed from any confined space if an upper limit ("ceiling") of 100 ppm is reached. Carbon monoxide exposure may lead to a significantly shorter life span due to heart damage. The carbon monoxide tolerance level for any person is altered by several factors, including activity level, rate of ventilation, a pre-existing cerebral or cardiovascular disease, cardiac output, anemia, sickle cell disease and other hematological disorders, barometric pressure, and metabolic rate.

The acute effects produced by carbon monoxide in relation to ambient concentration in parts per million are listed below:

Concentration	Symptoms
35 ppm (0.0035%)	Headache and dizziness within six to eight hours of constant exposure
100 ppm (0.01%)	Slight headache in two to three hours
200 ppm (0.02%)	Slight headache within two to three hours; loss of judgment
400 ppm (0.04%)	Frontal headache within one to two hours
800 ppm (0.08%)	Dizziness, nausea, and convulsions within 45 min; insensible within 2 hours
1,600 ppm (0.16%)	Headache, increased heart rate, dizziness, and nausea within 20 min; death in less than 2 hours
3,200 ppm (0.32%)	Headache, dizziness and nausea in five to ten minutes. Death within 30 minutes.
6,400 ppm (0.64%)	Headache and dizziness in one to two minutes. Convulsions, respiratory arrest, and death in less than 20 minutes.
12,800 ppm (1.28%)	Unconsciousness after 2–3 breaths. Death in less than three minutes.

Acute Poisoning

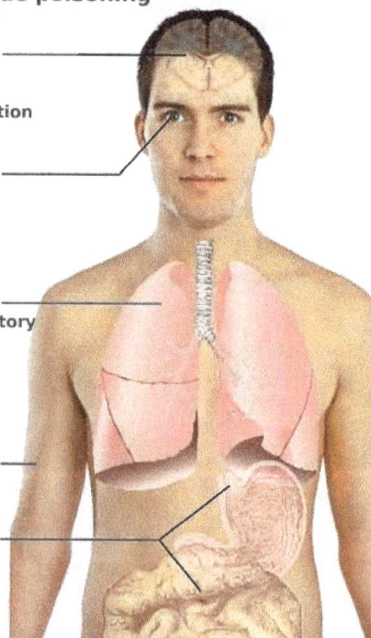

Symptoms of Carbon monoxide poisoning

- Dizziness
- Headache
- Disorientation
- Impairment of the cerebral function
- Coma

- Visual disturbances

- Disease of the heart and respiratory

- Muscle weakness
- Muscle cramps
- Seizures

- Nausea

- Aggravation of preexisting diseases

CO toxicity symptoms

The main manifestations of carbon monoxide poisoning develop in the organ systems most dependent on oxygen use, the central nervous system and the heart. The initial symptoms of acute carbon monoxide poisoning include headache, nausea, malaise, and fatigue. These symptoms are often mistaken for a virus such as influenza or other illnesses such as food poisoning or gastroenteritis. Headache is the most common symptom of acute carbon monoxide poisoning; it is often described as dull, frontal, and continuous. Increasing exposure produces cardiac abnormalities including fast heart rate, low blood pressure, and cardiac arrhythmia; central nervous system symptoms include delirium, hallucinations, dizziness, unsteady gait, confusion, seizures, central nervous system depression, unconsciousness, respiratory arrest, and death. Less common symptoms of acute carbon monoxide poisoning include myocardial ischemia, atrial fibrillation, pneumonia, pulmonary edema, high blood sugar, lactic acidosis, muscle necrosis, acute kidney failure, skin lesions, and visual and auditory problems.

One of the major concerns following acute carbon monoxide poisoning is the severe delayed neurological manifestations that may occur. Problems may include difficulty with higher intellectual functions, short-term memory loss, dementia, amnesia, psychosis, irritability, a strange gait, speech disturbances, Parkinson's disease-like syndromes, cortical blindness, and a depressed mood. Depression may occur in those who did not have pre-existing depression. These delayed neurological sequelae may occur in up to 50% of poisoned people after 2 to 40 days. It is difficult to predict who will develop delayed sequelae; however, advanced age, loss of consciousness while poisoned, and initial neurological abnormalities may increase the chance of developing delayed symptoms.

One classic sign of carbon monoxide poisoning is more often seen in the dead rather than the living – people have been described as looking red-cheeked and healthy. However, since this "cherry-red" appearance is common only in the deceased, and is unusual in living people, it is not considered a useful diagnostic sign in clinical medicine. In pathological (autopsy) examination the ruddy appearance of carbon monoxide poisoning is notable because unembalmed dead persons are normally bluish and pale, whereas dead carbon-monoxide poisoned persons may simply appear unusually lifelike in coloration. The colorant effect of carbon monoxide in such postmortem circumstances is thus analogous to its use as a red colorant in the commercial meat-packing industry.

Chronic Poisoning

Chronic exposure to relatively low levels of carbon monoxide may cause persistent headaches, lightheadedness, depression, confusion, memory loss, nausea and vomiting. It is unknown whether low-level chronic exposure may cause permanent neurological damage. Typically, upon removal from exposure to carbon monoxide, symptoms usually resolve themselves, unless there has been an episode of severe acute poisoning. However, one case noted permanent memory loss and learning problems after a 3-year exposure to relatively low levels of carbon monoxide from a faulty furnace. Chronic exposure may worsen cardiovascular symptoms in some people. Chronic carbon monoxide exposure might increase the risk of developing atherosclerosis. Long-term exposures to carbon monoxide present the greatest risk to persons with coronary heart disease and in females who are pregnant.

Causes

Concentration	Source
0.1 ppm	Natural atmosphere level (MOPITT)
0.5 to 5 ppm	Average level in homes
5 to 15 ppm	Near properly adjusted gas stoves in homes
100 to 200 ppm	Exhaust from automobiles in the Mexico City central area
5,000 ppm	Exhaust from a home wood fire
7,000 ppm	Undiluted warm car exhaust without a catalytic converter
30,000 ppm	Afterdamp following an explosion in a coal mine

Carbon monoxide is a product of combustion of organic matter under conditions of restricted oxygen supply, which prevents complete oxidation to carbon dioxide (CO_2). Sources of carbon monoxide include cigarette smoke, house fires, faulty furnaces, heaters, wood-burning stoves, internal combustion vehicle exhaust, electrical generators, propane-fueled equipment such as portable stoves, and gasoline-powered tools such as leaf blowers, lawn mowers, high-pressure washers, concrete cutting saws, power trowels, and welders. Exposure typically occurs when equipment is used in buildings or semi-enclosed spaces.

Riding in pickup trucks has led to poisoning in children. Idling automobiles with the exhaust pipe blocked by snow has led to the poisoning of car occupants. Any perforation between the exhaust manifold and shroud can result in exhaust gases reaching the cabin. Generators and propulsion engines on boats, especially houseboats, has resulted in fatal carbon monoxide exposures.

Poisoning may also occur following the use of a self-contained underwater breathing apparatus (SCUBA) due to faulty diving air compressors.

In caves carbon monoxide can build up in enclosed chambers due to the presence of decomposing organic matter. In coal mines incomplete combustion may occur during explosions resulting in the production of afterdamp. The gas is up to 3% CO and may be fatal after just a single breath. Following an explosion in a colliery adjacent, interconnected, mines may become dangerous due to the afterdamp leaking from mine to mine. Such an incident followed the Trimdon Grange explosion which killed men in the Kelloe mine.

Another source of poisoning is exposure to the organic solvent dichloromethane, found in some paint strippers, as the metabolism of dichloromethane produces carbon monoxide.

Pathophysiology

The precise mechanisms by which the effects of carbon monoxide are induced upon bodily systems, are complex and not yet fully understood. Known mechanisms include carbon monoxide binding to hemoglobin, myoglobin and mitochondrial cytochrome oxidase and restricting oxygen supply, and carbon monoxide causing brain lipid peroxidation.

Hemoglobin

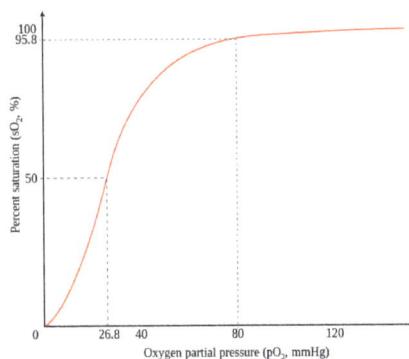

Carbon monoxide shifts the oxygen-dissociation curve to the left.

Carbon monoxide has a higher diffusion coefficient compared to oxygen and the only enzyme in the human body that produces carbon monoxide is heme oxygenase which is located in all cells and breaks down heme. Under normal conditions carbon monoxide levels in the plasma are approximately 0 mmHg because it has a higher diffusion coefficient and the body easily gets rid of any CO made. When CO is not ventilated it binds to hemoglobin, which is the principal oxygen-carrying compound in blood; this produces a compound known as carboxyhemoglobin. The traditional belief is that carbon monoxide toxicity arises from the formation of carboxyhemoglobin, which decreases the oxygen-carrying capacity of the blood and inhibits the transport, delivery, and utilization of oxygen by the body. The affinity between hemoglobin and carbon monoxide is approximately 230 times stronger than the affinity between hemoglobin and oxygen so hemoglobin binds to carbon monoxide in preference to oxygen.

Hemoglobin is a tetramer with four oxygen binding sites. The binding of carbon monoxide at one of these sites increases the oxygen affinity of the remaining three sites, which causes the hemoglobin molecule to retain oxygen that would otherwise be delivered to the tissue. This situation is described as carbon monoxide shifting the oxygen dissociation curve to the left. Because of the increased affinity between hemoglobin and oxygen during carbon monoxide poisoning, little oxygen will actually be released in the tissues. This causes hypoxic tissue injury. Hemoglobin acquires a bright red color when converted into carboxyhemoglobin, so poisoned cadavers and even commercial meats treated with carbon monoxide acquire an unnatural reddish hue.

Myoglobin

Carbon monoxide also binds to the hemeprotein myoglobin. It has a high affinity for myoglobin, about 60 times greater than that of oxygen. Carbon monoxide bound to myoglobin may impair its ability to utilize oxygen. This causes reduced cardiac output and hypotension, which may result in brain ischemia. A delayed return of symptoms have been reported. This results following a recurrence of increased carboxyhemoglobin levels; this effect may be due to a late release of carbon monoxide from myoglobin, which subsequently binds to hemoglobin.

Cytochrome Oxidase

Another mechanism involves effects on the mitochondrial respiratory enzyme chain that is responsible for effective tissue utilization of oxygen. Carbon monoxide binds to cytochrome oxidase with less affinity than oxygen, so it is possible that it requires significant intracellular hypoxia before binding. This binding interferes with aerobic metabolism and efficient adenosine triphosphate synthesis. Cells respond by switching to anaerobic metabolism, causing anoxia, lactic acidosis, and eventual cell death. The rate of dissociation between carbon monoxide and cytochrome oxidase is slow, causing a relatively prolonged impairment of oxidative metabolism.

Central Nervous System Effects

The mechanism that is thought to have a significant influence on delayed effects involves formed blood cells and chemical mediators, which cause brain lipid peroxidation (degradation of unsaturated fatty acids). Carbon monoxide causes endothelial cell and platelet release of nitric oxide, and the formation of oxygen free radicals including peroxynitrite. In the brain this causes further mitochondrial dysfunction, capillary leakage, leukocyte sequestration, and apoptosis. The result of these effects is lipid peroxidation, which causes delayed reversible demyelinization of white matter in the central nervous system known as Grinker myelinopathy, which can lead to edema and necrosis within the brain. This brain damage occurs mainly during the recovery period. This may result in cognitive defects, especially affecting memory and learning, and movement disorders. These disorders are typically related to damage to the cerebral white matter and basal ganglia. Hallmark pathological changes following poisoning are bilateral necrosis of the white matter, globus pallidus, cerebellum, hippocampus and the cerebral cortex.

Pregnancy

Carbon monoxide poisoning in pregnant women may cause severe adverse fetal effects. Poisoning causes fetal tissue hypoxia by decreasing the release of maternal oxygen to the fetus. Carbon monoxide also crosses the placenta and combines with fetal hemoglobin, causing more direct fetal tissue hypoxia. Additionally, fetal hemoglobin has a 10 to 15% higher affinity for carbon monoxide than adult hemoglobin, causing more severe poisoning in the fetus than in the adult. Elimination of carbon monoxide is slower in the fetus, leading to an accumulation of the toxic chemical. The level of fetal morbidity and mortality in acute carbon monoxide poisoning is significant, so despite mild maternal poisoning or following maternal recovery, severe fetal poisoning or death may still occur.

Diagnosis

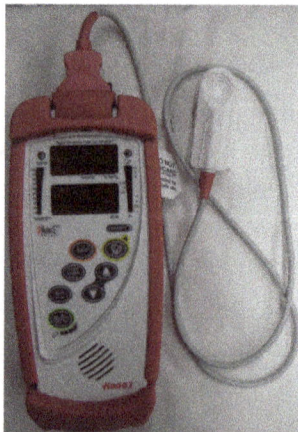

Finger tip carboxyhemoglobin saturation monitor (SpCO%). Note: This is not the same as a pulse oximeter (SpO2%), although some models (such as this one) do measure both the oxygen and carbon monoxide saturation.

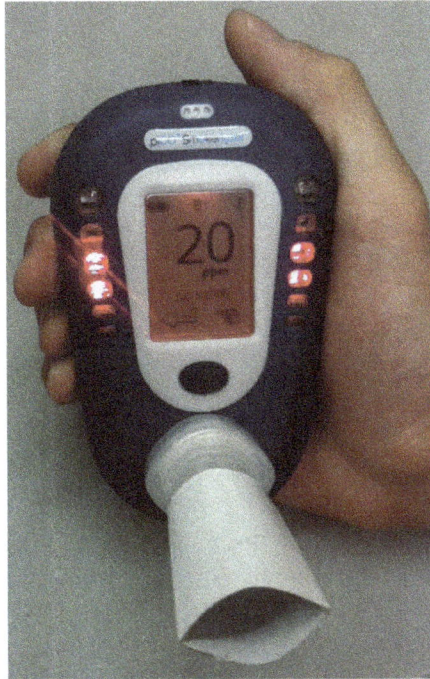

Breath CO monitor displaying carbon monoxide concentration of an exhaled breath sample (in ppm) with its corresponding percent concentration of carboxyhemoglobin.

As many symptoms of carbon monoxide poisoning also occur with many other types of poisonings and infections (such as the flu), the diagnosis is often difficult. A history of potential carbon monoxide exposure, such as being exposed to a residential fire, may suggest poisoning, but the diagnosis is confirmed by measuring the levels of carbon monoxide in the blood. This can be determined by measuring the amount of carboxyhemoglobin compared to the amount of hemoglobin in the blood.

As people may continue to experience significant symptoms of CO poisoning long after their blood carboxyhemoglobin concentration has returned to normal, presenting to examination with a normal carboxyhemoglobin level (which may happen in late states of poisoning) does not rule out poisoning.

A CO-oximeter is used to determine carboxyhemoglobin levels. Pulse CO-oximeters estimate carboxyhemoglobin with a non-invasive finger clip similar to a pulse oximeter. These devices function by passing various wavelengths of light through the fingertip and measuring the light absorption of the different types of hemoglobin in the capillaries.

The use of a regular pulse oximeter is not effective in the diagnosis of carbon monoxide poisoning as people suffering from carbon monoxide poisoning may have a normal oxygen saturation level on a pulse oximeter. This is due to the carboxyhemoglobin being misrepresented as oxyhemoglobin.

Breath CO monitoring offers a viable alternative to pulse CO-oximetry. Carboxyhemoglobin levels have been shown to have a strong correlation with breath CO concentration. However, many of these devices require the user to inhale deeply and hold their breath to allow the CO in the blood to escape into the lung before the measurement can be made. As this is not possible in a nonresponsive patient, these devices are not appropriate for use in on-scene emergency care detection of CO poisoning.

Detection in Biological Specimens

Carbon monoxide may be quantitated in blood using spectrophotometric methods or chromatographic techniques in order to confirm a diagnosis of poisoning in a person or to assist in the forensic investigation of a case of fatal exposure. Carboxyhemoglobin blood saturations may range up to 8–10% in heavy smokers or persons extensively exposed to automotive exhaust gases. In symptomatic poisoned people they are often in the 10–30% range, while persons who succumb may have postmortem blood levels of 30–90%.

The ratio of carboxyhemoglobin to hemoglobin molecules in an average person may be up to 5%, although cigarette smokers who smoke two packs/day may have levels up to 9%.

Differential Diagnosis

There are many conditions to be considered in the differential diagnosis of carbon monoxide poisoning. The earliest symptoms, especially from low level exposures, are often non-specific and readily confused with other illnesses, typically flu-like viral syndromes, depression, chronic fatigue syndrome, chest pain, and migraine or other headaches. Carbon monoxide has been called a "great mimicker" due to the presentation of poisoning being diverse and nonspecific. Other conditions included in the differential diagnosis include acute respiratory distress syndrome, altitude sickness, lactic acidosis, diabetic ketoacidosis, meningitis, methemoglobinemia, or opioid or toxic alcohol poisoning.

Prevention

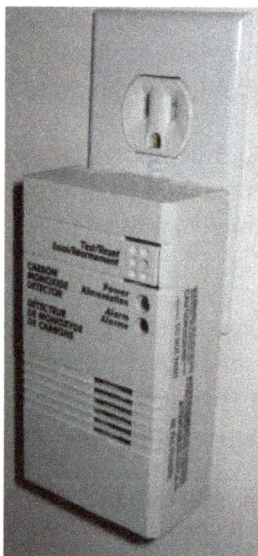

Carbon monoxide detector connected to a North American power outlet

Detectors

Prevention remains a vital public health issue, requiring public education on the safe operation of appliances, heaters, fireplaces, and internal-combustion engines, as well as increased emphasis on the installation of carbon monoxide detectors. Carbon monoxide is tasteless and odourless so can not be detected by smell.

The United States Consumer Product Safety Commission has stated, "carbon monoxide detectors are as important to home safety as smoke detectors are," and recommends each home have at least one carbon monoxide detector, and preferably one on each level of the building. These devices, which are relatively inexpensive and widely available, are either battery- or AC-powered, with or without battery backup. In buildings, carbon monoxide detectors are usually installed around heaters and other equipment. If a relatively high level of carbon monoxide is detected, the device sounds an alarm, giving people the chance to evacuate and ventilate the building. Unlike smoke detectors, carbon monoxide detectors do not need to be placed near ceiling level.

The use of carbon monoxide detectors has been standardized in many areas. In the USA, NFPA 720-2009, the carbon monoxide detector guidelines published by the National Fire Protection Association, mandates the placement of carbon monoxide detectors/alarms on every level of the residence, including the basement, in addition to outside sleeping areas. In new homes, AC-powered detectors must have battery backup and be interconnected to ensure early warning of occupants at all levels. NFPA 720-2009 is the first national carbon monoxide standard to address devices in non-residential buildings. These guidelines, which now pertain to schools, healthcare centers, nursing homes and other non-residential buildings, include three main points:

1. A secondary power supply (battery backup) must operate all carbon monoxide notification appliances for at least 12 hours,

2. Detectors must be on the ceiling in the same room as permanently installed fuel-burning appliances, and

3. Detectors must be located on every habitable level and in every HVAC zone of the building.

Gas organizations will often recommend to get gas appliances serviced at least once a year.

Legal Requirements

The NFPA standard is not necessarily enforced by law. As of April 2006, the U.S. state of Massachusetts requires detectors to be present in all residences with potential CO sources, regardless of building age and whether they are owner-occupied or rented. This is enforced by municipal inspectors, and was inspired by the death of 7-year-old Nicole Garofalo in 2005 due to snow blocking a home heating vent. Other jurisdictions may have no requirement or only mandate detectors for new construction or at time of sale.

Despite similar deaths in vehicles with clogged exhaust pipes (for example in the Northeastern United States blizzard of 1978 and February 2013 nor'easter) and the commercial availability of the equipment, there is no legal requirement for automotive CO detectors.

World Health Organization Recommendations

The following guideline values (ppm values rounded) and periods of time-weighted average exposures have been determined in such a way that the carboxyhaemoglobin (COHb) level of 2.5% is not exceeded, even when a normal subject engages in light or moderate exercise:

- 100 mg/m3 (87 ppm) for 15 min

- 60 mg/m3 (52 ppm) for 30 min

- 30 mg/m3 (26 ppm) for 1 h

- 10 mg/m3 (9 ppm) for 8 h

For indoor air quality 7 mg/m3 (6 ppm) for 24 h (so as not to exceed 2% COHb for chronic exposure)

Treatment

Initial treatment for carbon monoxide poisoning is to immediately remove the person from the exposure without endangering further people. Those who are unconscious may require CPR on site. Administering oxygen via non-rebreather mask shortens the half life of carbon monoxide from 320 minutes to 80 minutes on normal air. Oxygen hastens the dissociation of carbon monoxide from carboxyhemoglobin, thus turning it back into hemoglobin. Due to the possible severe effects in the fetus, pregnant women are treated with oxygen for longer periods of time than non-pregnant people.

Hyperbaric Oxygen

Hyperbaric oxygen is also used in the treatment of carbon monoxide poisoning, as it may hasten dissociation of CO from carboxyhemoglobin and cytochrome oxidase to a greater extent than normal oxygen. Hyperbaric oxygen at three times atmospheric pressure reduces the half life of carbon monoxide to 23 (~80/3 minutes) minutes, compared to 80 minutes for regular oxygen. It may also enhance oxygen transport to the tissues by plasma, partially bypassing the normal transfer through hemoglobin. However, it is controversial whether hyperbaric oxygen actually offers any extra benefits over normal high flow oxygen, in terms of increased survival or improved long-term outcomes.[needs update] There have been randomized controlled trials in which the two treatment options have been compared; of the six performed, four found hyperbaric oxygen improved outcome and two found no benefit for hyperbaric oxygen. Some of these trials have been criticized for apparent flaws in their implementation. A review of all the literature on carbon monoxide poisoning treatment concluded that the role of hyperbaric oxygen is unclear and the available evidence neither confirms nor denies a medically meaningful benefit. The authors suggested a large, well designed, externally audited, multicentre trial to compare normal oxygen with hyperbaric oxygen.

Other

Further treatment for other complications such as seizure, hypotension, cardiac abnormalities, pulmonary edema, and acidosis may be required. Increased muscle activity and seizures should be treated with dantrolene or diazepam; diazepam should only be given with appropriate respiratory support. Hypotension requires treatment with intravenous fluids; vasopressors may be required to treat myocardial depression. Cardiac dysrhythmias are treated with standard advanced cardiac life support protocols. If severe, metabolic acidosis is treated with sodium bicarbonate. Treatment with sodium bicarbonate is controversial as acidosis may increase tissue oxygen availability. Treatment of acidosis may only need to consist of oxygen therapy. The delayed development of neuropsychiatric impairment is one of the most serious complications of carbon monoxide poisoning. Brain damage is confirmed following MRI or CAT scans. Extensive follow up and supportive treatment is often required for delayed neurological damage. Outcomes are often difficult to predict following poisoning, especially people who have symptoms of cardiac arrest, coma, metabolic acidosis, or have high carboxyhemoglobin levels. One study reported that approximately 30% of people with severe carbon monoxide poisoning will have a fatal outcome. It has been reported that electroconvulsive therapy (ECT) may increase the likelihood of delayed neuropsychiatric sequelae (DNS) after carbon monoxide (CO) poisoning.

Epidemiology

The true number of incidents of carbon monoxide poisoning is unknown, since many non-lethal exposures go undetected. From the available data, carbon monoxide poisoning is the most common cause of injury and death due to poisoning worldwide. Poisoning is typically more common during the winter months. This is due to increased domestic use of gas furnaces, gas or kerosene space heaters, and kitchen stoves during the winter months, which if faulty and/or used without adequate ventilation, may produce excessive carbon monoxide. Carbon monoxide detection and poisoning also increases during power outages.

It has been estimated that more than 40,000 people per year seek medical attention for carbon monoxide poisoning in the United States. 95% of carbon monoxide poisoning deaths in the United States are due to gas space heaters. In many industrialized countries carbon monoxide is the cause of more than 50% of fatal poisonings. In the United States, approximately 200 people die each year from carbon monoxide poisoning associated with home fuel-burning heating equipment. Carbon monoxide poisoning contributes to the approximately 5613 smoke inhalation deaths each year in the United States. The CDC reports, "Each year, more than 500 Americans die from unintentional carbon monoxide poisoning, and more than 2,000 commit suicide by intentionally poisoning themselves." For the 10-year period from 1979 to 1988, 56,133 deaths from carbon monoxide poisoning occurred in the United States, with 25,889 of those being suicides, leaving 30,244 unintentional deaths. A report from New Zealand showed that 206 people died from carbon monoxide poisoning in the years of 2001 and 2002. In total carbon monoxide poisoning was responsible for 43.9% of deaths by poisoning in that country. In South Korea, 1,950 people had been poisoned by carbon monoxide with 254 deaths from 2001 through 2003. A report from Jerusalem showed 3.53 per 100,000 people were poisoned annually from 2001 through 2006. In Hubei, China, 218 deaths from poisoning were reported over a 10-year period with 16.5% being from carbon monoxide exposure.

Society and Culture

As part of the Holocaust during World War II, German Nazis used gas vans at Chelmno camp and elsewhere to kill an estimated over 700,000 prisoners by carbon monoxide poisoning. This method was also used in the gas chambers of several death camps such as Treblinka, Sobibor and Belzec. Gassing with carbon monoxide started in action T4, the euthanasia programme developed by the Nazis in Germany to murder the mentally ill and disabled people before the war started in earnest. Many key personnel were recruited to murder much larger numbers of people in the gas vans and the special gas chambers used in the death camps such as Treblinka. Exhaust fumes from tank engines for example, were used to supply the gas to the chambers.

Research

Carbon monoxide is produced naturally by the body as a byproduct of converting protoporphyrin into bilirubin. This carbon monoxide also combines with hemoglobin to make carboxyhemoglobin, but not at toxic levels.

Small amounts of CO are beneficial and enzymes exist that produce it at times of oxidative stress. Drugs are being developed to introduce small amounts of CO during certain kinds of surgery, these drugs are called Carbon monoxide-releasing molecules.

Polycyclic Aromatic Hydrocarbon

Polycyclic aromatic hydrocarbons (PAHs, also *polyaromatic hydrocarbons*) are hydrocarbons—organic compounds containing only carbon and hydrogen—that are composed of multiple aromatic rings (organic rings in which the electrons are delocalized). Formally, the class is further defined as lacking further branching substituents on these ring structures. Polynuclear aromatic hydrocarbons (PNAs) are a subset of PAHs that have fused aromatic rings, that is, rings that share one or more sides. The simplest such chemicals are naphthalene, having two aromatic rings, and the three-ring compounds anthracene and phenanthrene.

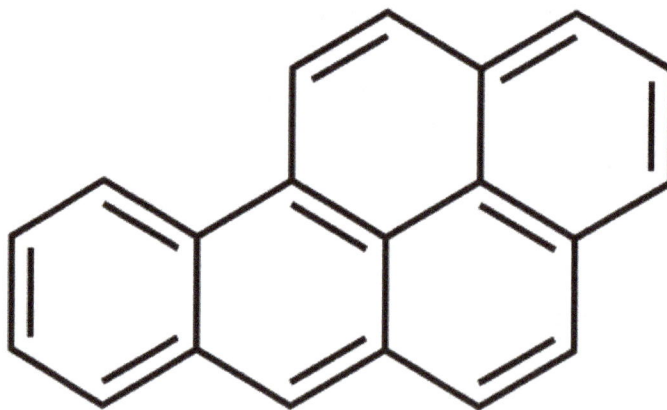

Standard line-angle schematic representation of an important PAH, benzo[*a*]pyrene, where carbon atoms are represented by the vertices of the hexagons, and hydrogens are inferred as projecting out at 120° angles to fill the fourth carbon valence (as necessary)

PAHs are neutral, nonpolar molecules found in coal and in tar deposits. They are produced as well by incomplete combustion of organic matter (e.g., in engines and incinerators, when biomass burns in forest fires, etc.).

PAHs may also be abundant in the universe, and are conjectured to have formed as early as the first couple of billion years after the Big Bang, in association with formation of new stars and exoplanets. Some studies suggest that PAHs account for a significant percentage of all carbon in the universe, and PAHs are discussed as possible starting materials for abiologic syntheses of materials required by the earliest forms of life.

Nomenclature, Structure, Properties

Nomenclature and Structure

The tricyclic species phenanthrene and anthracene represent the starting members of the PAHs. Smaller molecules, such as benzene, are not PAHs, and PAHs are not generally considered to contain heteroatoms or carry substituents.

PAHs with five or six-membered rings are most common. Those composed only of six-membered rings are called *alternant* PAHs, which include *benzenoid* PAHs. The following are examples of PAHs that vary in the number and arrangement of their rings:

- Examples of PAH compounds

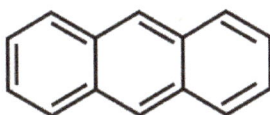

Anthracene

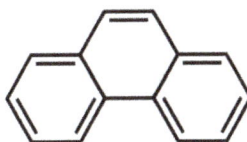

Phenanthrene

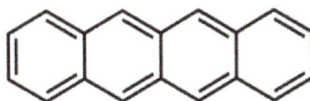

Tetracene

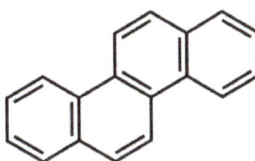

Chrysene

Triphenylene

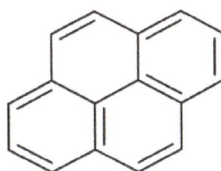

Pyrene

Physicochemical Properties and Bonding

PAHs are nonpolar and lipophilic. PAHs are insoluble in water. The larger members are also poorly soluble in organic solvents as well as lipids. They are usually colorless.

Although PAHs clearly are aromatic compounds, the degree of aromaticity can be different for each ring segment. According to *Clar's rule* (formulated by Erich Clar in 1964) for PAHs the resonance structure with the largest number of disjoint aromatic π-sextets—i.e. benzene-like moieties—is the most important for the characterization of the properties.

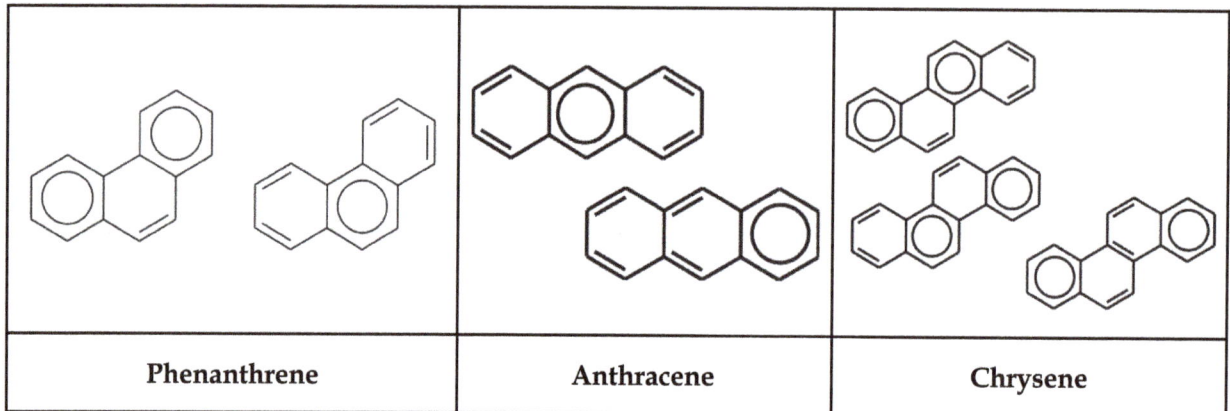

| Phenanthrene | Anthracene | Chrysene |

For example, in phenanthrene one Clar structure has two sextets at the extremities, while the other resonance structure has just one central sextet; therefore in this molecule the outer rings have greater aromatic character whereas the central ring is less aromatic and therefore more reactive. In contrast, in anthracene the resonance structures have one sextet, which can be at any of the three rings, and the aromaticity spreads out more evenly across the whole molecule. This difference in number of sextets is reflected in the UV absorbance spectra of these two isomers; phenanthrene has a highest wavelength absorbance around 290 nm, while anthracene has highest wavelength bands around 380 nm. Three Clar structures with two sextets each are present in chrysene. Superposition of these structures reveals that the aromaticity in the outer rings is greater (each has a sextet in two of the three Clar structures) compared to the inner rings (each has a sextet in only one of the three).

Sources & Distribution

Polycyclic aromatic hydrocarbons are primarily found in natural sources such as creosote. They can result from the incomplete combustion of organic matter. PAHs can also be produced geologically when organic sediments are chemically transformed into fossil fuels such as oil and coal. The dominant sources of PAHs in the environment are thus from human activity: Wood-burning and combustion of other biofuels such as dung or crop residues contribute more than half of annual global PAH emissions, particularly due to biofuel use in India and China. As of 2004, industrial processes and the extraction and use of fossil fuels made up slightly more than one quarter of global PAH emissions, dominating outputs in industrial countries such as the United States. Wild fires are another notable source. Substantially higher outdoor air, soil, and water concentrations of PAHs have been measured in Asia, Africa, and Latin America than in Europe, Australia, and the U.S./Canada.

PAHs are typically found as complex mixtures. Lower-temperature combustion, such as tobacco smoking or wood-burning, tends to generate low molecular weight PAHs, whereas high-temperature industrial processes typically generate PAHs with higher molecular weights.

In the Aqueous Environment

Most PAH's are insoluble in water, which limits their mobility in the environment. Aqueous solubility of PAHs decreases approximately logarithmically as molecular mass increases.

Two-ring PAHs, and to a lesser extent three-ring PAHs, dissolve in water, making them more available for biological uptake and degradation. Further, two- to four-ring PAHs volatilize sufficiently to appear in the atmosphere predominantly in gaseous form, although the physical state of four-ring PAHs can depend on temperature. In contrast, compounds with five or more rings have low solubility in water and low volatility; they are therefore predominantly in solid state, bound to particulate air pollution, soils, or sediments. In solid state, these compounds are less accessible for biological uptake or degradation, increasing their persistence in the environment.

Human Exposure

Human exposure varies across the globe and depends on factors such as smoking rates, fuel types in cooking, and pollution controls on power plants, industrial processes, and vehicles. Developed countries with stricter air and water pollution controls, cleaner sources of cooking (i.e., gas and electricity vs. coal or biofuels), and prohibitions of public smoking tend to have lower levels of PAH exposure, while developing and undeveloped countries tend to have higher levels.

A wood-burning open-air cook stove. Smoke from solid fuels like wood is a large source of PAHs globally.

Burning solid fuels such as coal and biofuels in the home for cooking and heating is a dominant global source of PAH emissions that in developing countries leads to high levels of exposure to indoor particulate air pollution containing PAHs, particularly for women and children who spend more time in the home or cooking.

In industrial countries, people who smoke tobacco products, or who are exposed to second-hand smoke, are among the most highly exposed groups; tobacco smoke contributes to 90% of indoor PAH levels in the homes of smokers. For the general population in developed countries, the diet is otherwise the dominant source of PAH exposure, particularly from smoking or grilling meat or consuming PAHs deposited on plant foods, especially broad-leafed vegetables, during growth. PAHs are typically at low concentrations in drinking water.

Smog in Cairo, Egypt. Particulate air pollution, including smog, is a substantial avenue for human exposure to PAHs.

Emissions from vehicles such as cars and trucks can be a substantial outdoor source of PAHs in particulate air pollution. Geographically, major roadways are thus sources of PAHs, which may distribute in the atmosphere or deposit nearby. Catalytic converters are estimated to reduce PAH emissions from gasoline-fired vehicles by 25-fold.

People can also be occupationally exposed during work that involves fossil fuels or their derivatives, wood burning, carbon electrodes, or exposure to diesel exhaust. Industrial activity that can produce and distribute PAHs includes aluminum, iron, and steel manufacturing; coal gasification, tar distillation, shale oil extraction; production of coke, creosote, carbon black, and calcium carbide; road paving and asphalt manufacturing; rubber tire production; manufacturing or use of metal working fluids; and activity of coal or natural gas power stations.

Environmental distribution and Degradation

PAHs typically disperse from urban and suburban non-point sources through road run-off, sewage, and atmospheric circulation and subsequent deposition of particulate air pollution. Soil and river sediment near industrial sites such as creosote manufacturing facilities can be highly contaminated with PAHs. Oil spills, creosote, coal mining dust, and other fossil fuel sources can also distribute PAHs in the environment.

Oil on a beach after a 2007 oil spill in Korea.

Two- and three-ring PAHs can disperse widely while dissolved in water or as gases in the atmosphere, while PAHs with higher molecular weights can disperse locally or regionally adhered to particulate matter that is suspended in air or water until the particles land or settle out of the water column. PAHs have a strong affinity for organic carbon, and thus highly organic sediments in rivers, lakes, and the ocean can be a substantial sink for PAHs.

Algae and some invertebrates such as protozoans, mollusks, and many polychaetes have limited ability to metabolize PAHs and bioaccumulate disproportionate concentrations of PAHs in their tissues; however, PAH metabolism can vary substantially across invertebrate species. Most vertebrates metabolize and excrete PAHs relatively rapidly. Tissue concentrations of PAHs do not increase (biomagnify) from the lowest to highest levels of food chains.

PAHs transform slowly to a wide range of degradation products. Biological degradation by microbes is a dominant form of PAH transformation in the environment. Soil-consuming invertebrates such as earthworms speed PAH degradation, either through direct metabolism or by improving the conditions for microbial transformations. Abiotic degradation in the atmosphere and the top layers of surface waters can produce nitrogenated, halogenated, hydroxylated, and oxygenated PAHs; some of these compounds can be more toxic, water-soluble, and mobile than their parent PAHs.

Minor Sources

Volcanic eruptions may emit PAHs. Certain PAHs such as perylene can also be generated in anaerobic sediments from existing organic material, although it remains undetermined whether abiotic or microbial processes drive their production.

Human Health

Cancer is a primary human health risk of exposure to PAHs. Exposure to PAHs has also been linked with cardiovascular disease and poor fetal development.

Cancer

PAHs have been linked to skin, lung, bladder, liver, and stomach cancers in well-established animal model studies. Specific compounds classified by various agencies as possible or probable human carcinogens are identified in the section "Regulation and Oversight" below.

Historical Significance

An eighteenth century drawing of chimney sweeps.

Historically, PAHs contributed substantially to our understanding of adverse health effects from exposures to environmental contaminants, including chemical carcinogenesis. In 1775, Percivall Pott, a surgeon at St. Bartholomew's Hospital in London, observed that scrotal cancer was unusually common in chimney sweepers and proposed the cause as occupational exposure to soot. A century later, Richard von Volkmann reported increased skin cancers in workers of the coal tar industry of Germany, and by the early 1900s increased rates of cancer from exposure to soot and coal tar was widely accepted. In 1915, Yamigawa and Ichicawa were the first to experimentally produce cancers, specifically of the skin, by topically applying coal tar to rabbit ears.

In 1922, Ernest Kennaway determined that the carcinogenic component of coal tar mixtures was an organic compound consisting of only C and H. This component was later linked to a characteristic fluorescent pattern that was similar but not identical to benz[*a*]anthracene, a PAH that was subsequently demonstrated to cause tumors. Cook, Hewett and Hieger then linked the specific spectroscopic fluorescent profile of benzo[*a*]pyrene to that of the carcinogenic component of coal tar, the first time that a specific compound from an environmental mixture (coal tar) was demonstrated to be carcinogenic.

In the 1930s and later, epidemiologists from Japan, England, and the U.S., including Richard Doll and various others, reported greater rates of death from lung cancer following occupational exposure to PAH-rich environments among workers in coke ovens and coal carbonization and gasification processes.

Mechanisms of Carcinogenesis

The structure of a PAH influences whether and how the individual compound is carcinogenic. Some carcinogenic PAHs are genotoxic and induce mutations that initiate cancer; others are not genotoxic and instead affect cancer promotion or progression.

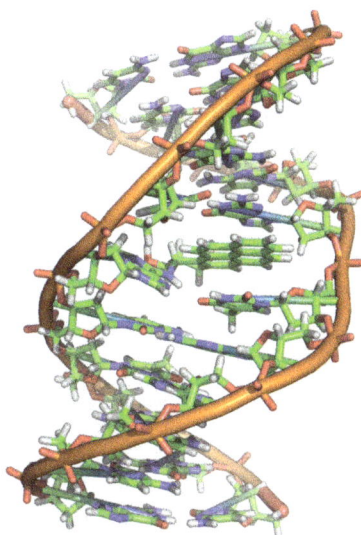

An adduct formed between a DNA strand and an epoxide derived from a benzo[*a*]pyrene molecule (center); such adducts may interfere with normal DNA replication.

PAHs that affect cancer initiation are typically first chemically modified by enzymes into metabolites that react with DNA, leading to mutations. When the DNA sequence is altered in genes that regulate cell replication, cancer can result. Mutagenic PAHs, such as benzo[*a*]pyrene, usually have four or more aromatic rings as well as a "bay region", a structural pocket that increases reactivity of the molecule to the metabolizing enzymes. Mutagenic metabolites of PAHs include diol epoxides, quinones, and radical PAH cations. These metabolites can bind to DNA at specific sites, forming bulky complexes called DNA adducts that can be stable or unstable. Stable adducts may lead to DNA replication errors, while unstable adducts react with the DNA strand, removing a purine base (either adenine or guanine). Such mutations, if they are not repaired, can transform genes encoding for normal cell signaling proteins into cancer-causing oncogenes. Quinones can also repeatedly generate reactive oxygen species that may independently damage DNA.

Enzymes in the cytochrome family (CYP1A1, CYP1A2, CYP1B1) metabolize PAHs to diol epoxides. PAH exposure can increase production of the cytochrome enzymes, allowing the enzymes to convert PAHs into mutagenic diol epoxides at greater rates. In this pathway, PAH molecules bind to the aryl hydrocarbon receptor (AhR) and activate it as a transcription factor that increases production of the cytochrome enzymes. The activity of these enzymes may at times conversely protect against PAH toxicity, which is not yet well understood.

Low molecular weight PAHs, with 2 to 4 aromatic hydrocarbon rings, are more potent as co-carcinogens during the promotional stage of cancer. In this stage, an initiated cell (i.e., a cell that has retained a carcinogenic mutation in a key gene related to cell replication) is removed from growth-suppressing signals from its neighboring cells and begins to clonally replicate. Low molecular weight PAHs that have bay or bay-like regions can dysregulate gap junction channels, interfering with intercellular communication, and also affect mitogen-activated protein kinases that activate transcription factors involved in cell proliferation. Closure of gap junction protein channels is a normal precursor to cell division. Excessive closure of these channels after exposure to PAHs results in removing a cell from the normal growth-regulating signals imposed by its local community of cells, thus allowing initiated cancerous cells to replicate. These PAHs do not need to be enzymatically metabolized first. Low molecular weight PAHs are prevalent in the environment, thus posing a significant risk to human health at the promotional phases of cancer.

Cardiovascular Disease

Adult exposure to PAHs has been linked to cardiovascular disease. PAHs are among the complex suite of contaminants in cigarette smoke and particulate air pollution and may contribute to cardiovascular disease resulting from such exposures.

In laboratory experiments, animals exposed to certain PAHs have shown increased development of plaques (atherogenesis) within arteries. Potential mechanisms for the pathogenesis and development of atherosclerotic plaques may be similar to the mechanisms involved in the carcinogenic and mutagenic properties of PAHs. A leading hypothesis is that PAHs may activate the cytochrome enzyme CYP1B1 in vascular smooth muscle cells. This enzyme then metabolically processes the PAHs to quinone metabolites that bind to DNA in reactive adducts that remove purine bases. The resulting mutations may contribute to unregulated growth of vascular smooth muscle cells or to their migration to the inside of the artery, which are steps in plaque formation. These quinone metabolites also generate reactive oxygen species that may alter the activity of genes that affect plaque formation.

Oxidative stress following PAH exposure could also result in cardiovascular disease by causing inflammation, which has been recognized as an important factor in the development of atherosclerosis and cardiovascular disease. Biomarkers of exposure to PAHs in humans have been associated with inflammatory biomarkers that are recognized as important predictors of cardiovascular disease, suggesting that oxidative stress resulting from exposure to PAHs may be a mechanism of cardiovascular disease in humans.

Developmental Impacts

Multiple epidemiological studies of people living in Europe, the United States, and China have linked in utero exposure to PAHs, through air pollution or parental occupational exposure, with poor fetal growth, reduced immune function, and poorer neurological development, including lower IQ.

Regulation and Oversight

Some governmental bodies, including the European Union as well as NIOSH and the Environmental Protection Agency in the US, regulate concentrations of PAHs in air, water, and soil. The European Commission has restricted concentrations of 8 carcinogenic PAHs in consumer products that contact the skin or mouth.

Priority polycyclic aromatic hydrocarbons identified by the US EPA, the US Agency for Toxic Substances and Disease Registry (ATSDR), and the European Food Safety Authority (EFSA) due to their carcinogenicity or genotoxicity and/or ability to be monitored are the following:

Compound	Agency	Compound	Agency
acenaphthene	EPA, ATSDR	cyclopenta[c,d]pyrene	EFSA
acenaphthylene	EPA, ATSDR	dibenz[a,h]anthracene[A]	EPA, ATSDR, EFSA
anthracene	EPA, ATSDR	dibenzo[a,e]pyrene	EFSA
benz[a]anthracene[A]	EPA, ATSDR, EFSA	dibenzo[a,h]pyrene	EFSA
benzo[b]fluoranthene[A]	EPA, ATSDR, EFSA	dibenzo[a,i]pyrene	EFSA
benzo[j]fluoranthene	ATSDR, EFSA	dibenzo[a,l]pyrene	EFSA
benzo[k]fluoranthene[A]	EPA, ATSDR, EFSA	fluoranthene	EPA, ATSDR
benzo[c]fluorene	EFSA	fluorene	EPA, ATSDR
benzo[g,h,i]perylene[A]	EPA, ATSDR, EFSA	indeno[1,2,3-c,d]pyrene[A]	EPA, ATSDR, EFSA
benzo[a]pyrene[A]	EPA, ATSDR, EFSA	5-methylchrysene	EFSA
benzo[e]pyrene	ATSDR	naphthalene	EPA
chrysene[A]	EPA, ATSDR, EFSA	phenanthrene	EPA, ATSDR
coronene	ATSDR	pyrene	EPA, ATSDR

A Considered probable or possible human carcinogens by the US EPA, the European Union, and/or the International Agency for Research on Cancer (IARC).

Detection and Optical Properties

A spectral database exists for tracking polycyclic aromatic hydrocarbons (PAHs) in the universe. greatly upgraded database Detection of PAHs in materials is often done using gas chromatography-mass spectrometry or liquid chromatography with ultraviolet-visible or fluorescence spectroscopic methods or by using rapid test PAH indicator strips.

PAHs possess very characteristic UV absorbance spectra. These often possess many absorbance bands and are unique for each ring structure. Thus, for a set of isomers, each isomer has a different UV absorbance spectrum than the others. This is particularly useful in the identification of PAHs. Most PAHs are also fluorescent, emitting characteristic wavelengths of light when they are excited (when the molecules absorb light). The extended pi-electron electronic structures of PAHs lead to these spectra, as well as to certain large PAHs also exhibiting semi-conducting and other behaviors.

Origins of Life

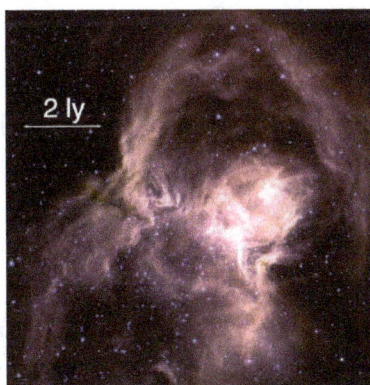

The Spitzer Space Telescope captures PAH spectral lines, producing this image of nebulosity in a stellar nursery.

PAHs may be abundant in the universe. They seem to have been formed as early as a couple of billion years after the Big Bang, and are associated with new stars and exoplanets. More than 20% of the carbon in the universe may be associated with PAHs. PAHs are considered possible starting material for the earliest forms of life. Light emitted by the Red Rectangle nebula and found spectral signatures that suggest the presence of anthracene and pyrene. This report was considered a controversial hypothesis that as nebulae of the same type as the Red Rectangle approach the ends of their lives, convection currents cause carbon and hydrogen in the nebulae's core to get caught in stellar winds, and radiate outward. As they cool, the atoms supposedly bond to each other in various ways and eventually form particles of a million or more atoms. Witt and his team inferred that PAHs —which may have been vital in the formation of early life on Earth— in a nebula, by necessity they must originate in nebulae.

Two extremely bright stars illuminate a mist of PAHs in this Spitzer image.

More recently, fullerenes (or "buckyballs"), have been detected in other nebulae. Fullerenes are also implicated in the origin of life; according to astronomer Letizia Stanghellini, "It's possible that buckyballs from outer space provided seeds for life on Earth." In September 2012, NASA scientists reported results of analog studies *in vitro* that PAHs, subjected to interstellar medium (ISM) conditions, are transformed, through hydrogenation, oxygenation, and hydroxylation, to more complex organics—"a step along the path toward amino acids and nucleotides, the raw materials of proteins and DNA, respectively". Further, as a result of these transformations, the PAHs lose their spectroscopic signature which could be one of the reasons "for the lack of PAH detection in interstellar ice grains, particularly the outer regions of cold, dense clouds or the upper molecular layers of protoplanetary disks."

PAHs have been detected in the upper atmosphere of Titan, the largest moon of the planet Saturn.

References

- Bánfalvi, G (2011). "Heavy Metals, Trace Elements and their Cellular Effects". In Bánfalvi, G. Cellular Effects of Heavy Metals. Springer. pp. 3–28. ISBN 9789400704275.

- Haines, AT; Nieboer, E (1988). "Chromium hypersensitivity". In Nriagu, JO; Nieboer, E. Chromium in the Natural and Human Environments. Wiley. pp. 497–532. ISBN 0471856436.

- Landis, WG; Sofield, RM; Yu, M-H (2000). Introduction to Environmental Toxicology: Molecular Substructures to Ecological Landscapes. 4th: CRC Press. ISBN 9781439804100.

- Lovei, M (1998). Phasing Out Lead from Gasoline: Worldwide Experience and Policy Implications. World Bank Technical Paper. 397. The World Bank. ISBN 082134157X. ISSN 0253-7494.

- Nielen, MWF; Marvin, HJP (2008). "Challenges in Chemical Food Contaminants and Residue Analysis". In Picó, Y. Food Contaminants and Residue Analysis. Elsevier. pp. 1–28. ISBN 0080931928.

- Pezzarossa, B; Gorini, F; Petruzelli, G (2011). "Heavy Metal and Selenium Distribution and Bioavailability in Contaminated Sites: A Tool for Phytoremediation". In Selim, HM. Dynamics and Bioavailabiliy of Heavy Metals in the Rootzone. CRC Press. pp. 93–128. ISBN 9781439826225.

- Rand, GM; Wells, PG; McCarty, LS (1995). "Introduction to aquatic toxicology". In Rand, GM. Fundamentals Of Aquatic Toxicology: Effects, Environmental Fate And Risk Assessment (2nd ed.). Taylor & Francis. pp. 3–70. ISBN 1560320907.

- Rogers, MJ (2000). "Text and Illustrations. Dioscorides and the Illuminated Herbal in the Arab Tradition". In Contadini, A. Arab Painting: Text and Image in Illustrated Arabic Manuscripts. Leiden: Koninklijke Brill NV. pp. 41–48 (41). ISBN 9789004186309.

- Sengupta, AK (2002). "Principles of Heavy Metals Separation". In Sengupta, AK. Environmental Separation of Heavy Metals: Engineering Processes. Lewis. ISBN 1566768845.

- Srivastava, S; Goyal, P (2010). Novel Biomaterials: Decontamination of Toxic Metals from Wastewater. Springer-Verlag. ISBN 978-3-642-11329-1.

- Collins MA, Brickle P, Brown J and Belchier M (2010) "The Patagonian toothfish: biology, ecology and fishery" In: M Lesser (Ed.) Advances in Marine Biology, Volume 58, pp. 229–289, Academic Press. ISBN 978-0-12-381015-1.

- Casarett, LJ; Klaassen, CD; Doull, J, eds. (2007). "Toxic effects of metals". Casarett and Doull's Toxicology: The Basic Science of Poisons (7th ed.). McGraw-Hill Professional. ISBN 0-07-147051-4.

- Chisolm, J.J. (2004). "Lead poisoning". In Crocetti, M.; Barone, M.A.; Oski, F.A. Oski's Essential Pediatrics (2nd ed.). Lippincott Williams & Wilkins. ISBN 0-7817-3770-2.

- Dart, R.C.; Hurlbut, K.M.; Boyer-Hassen, L.V. (2004). "Lead". In Dart, RC. Medical Toxicology (3rd ed.). Lippincott Williams & Wilkins. ISBN 0-7817-2845-2.

- Grant, L.D. (2009). "Lead and compounds". In Lippmann, M. Environmental Toxicants: Human Exposures and Their Health Effects (3rd ed.). Wiley-Interscience. ISBN 0-471-79335-3.

- Henretig F.M. (2006). "Lead". In Goldfrank, LR. Goldfrank's Toxicologic Emergencies (8th ed.). McGraw-Hill Professional. ISBN 0-07-143763-0.

Insecticides, Pesticides and Environmental Toxicology

The negative impact of using insecticides and pesticides on the environment has been well documented. Persistent organic pollutants like endrin, mirex, hexachlorobenzene and other such pesticides do not degrade easily and have really long half-lives that lead to their bioaccumulation. In this chapter, the reader is presented with information regarding the various pesticides and insecticides that damage the environment and are proven carcinogens and their impact on the flora and fauna.

Persistent Organic Pollutant

Persistent organic pollutants (POPs) are organic compounds that are resistant to environmental degradation through chemical, biological, and photolytic processes. Because of their persistence, POPs bioaccumulate with potential significant impacts on human health and the environment. The effect of POPs on human and environmental health was discussed, with intention to eliminate or severely restrict their production, by the international community at the Stockholm Convention on Persistent Organic Pollutants in 2001.

Many POPs are currently or were in the past used as pesticides, solvents, pharmaceuticals, and industrial chemicals. Although some POPs arise naturally, for example volcanoes and various biosynthetic pathways, most are man-made via total synthesis.

Consequences of Persistence

POPs typically are halogenated organic compounds and as such exhibit high lipid solubility. For this reason, they bioaccumulate in fatty tissues. Halogenated compounds also exhibit great stability reflecting the nonreactivity of C-Cl bonds toward hydrolysis and photolytic degradation. The stability and lipophilicity of organic compounds often correlates with their halogen content, thus polyhalogenated organic compounds are of particular concern. They exert their negative effects on the environment through two processes, long range transport, which allows them to travel far from their source, and bioaccumulation, which reconcentrates these chemical compounds to potentially dangerous levels. Compounds that make up POPs are also classed as PBTs (Persistent, Bioaccumulative and Toxic) or TOMPs (Toxic Organic Micro Pollutants).

Long-range Transport

POPs enter the gas phase under certain environmental temperatures and volatize from soils, vegetation, and bodies of water into the atmosphere, resisting breakdown reactions in the air, to travel long distances before being re-deposited. This results in accumulation of POPs in areas far from where they were used or emitted, specifically environments where POPs have never been introduced such as Antarctica, and the Arctic circle. POPs can be present as vapors in the atmosphere or bound to the surface of solid particles. POPs have low solubility in water but are easily captured by solid particles, and are soluble in organic fluids (oils, fats, and liquid fuels). POPs are not easily degraded in the environment due to their stability and low decomposition rates. Due to this capacity for long-range transport, POP environmental contamination is extensive, even in areas where POPs have never been used, and will remain in these environments years after restrictions implemented due to their resistance to degradation.

Bioaccumulation

Bioaccumulation of POPs is typically associated with the compounds high lipid solubility and ability to accumulate in the fatty tissues of living organisms for long periods of time. Persistent chemicals tend to have higher concentrations and are eliminated more slowly. Dietary accumulation or bioaccumulation is another hallmark characteristic of POPs, as POPs move up the food chain, they increase in concentration as they are processed and metabolized in certain tissues of organisms. The natural capacity for animals gastrointestinal tract concentrate ingested chemicals, along with poorly metabolized and hydrophobic nature of POPs makes such compounds highly susceptible to bioaccumulation. Thus POPs not only persist in the environment, but also as they are taken in by animals they bioaccumulate, increasing their concentration and toxicity in the environment.

Stockholm Convention on Persistent Organic Pollutants

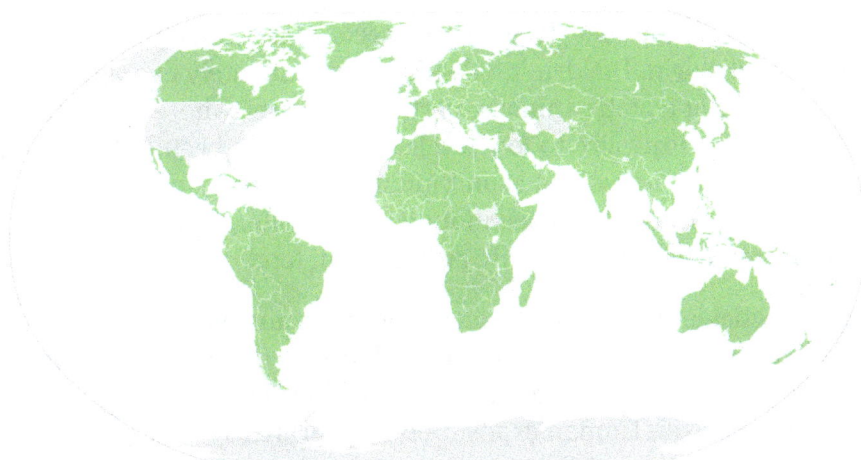

State parties to the Stockholm Convention on Persistent Organic Pollutants

The Stockholm Convention was adopted and put into practice by the United Nations Environment Programme (UNEP) on May 22, 2001. The UNEP decided that POP regulation needed to be addressed globally for the future. The purpose statement of the agreement is "to protect human health and the environment from persistent organic pollutants." As of 2014, there are 179 countries in compliance with the Stockholm convention. The convention and its participants have recognized the potential human and environmental toxicity of POPs. They recognize that POPs have the potential for long range transport and bioaccumulation and biomagnification. The convention seeks to study and then judge whether or not a number of chemicals that have been developed with advances in technology and science can be categorized as POPs or not. The initial meeting in 2001 made a preliminary list, termed the "dirty dozen," of chemicals that are classified as POPs. As of 2014, the United States of America has signed the Stockholm Convention but has not ratified it. There are a handful of other countries that have not ratified the convention but most countries in the world have ratified the convention.

Compounds on the Stockholm Convention list

In May 1995, the United Nations Environment Programme Governing Council investigated POPs. Initially the Convention recognized only twelve POPs for their adverse effects on human health and the environment, placing a global ban on these particularly harmful and toxic compounds and requiring its parties to take measures to eliminate or reduce the release of POPs in the environment.

1. Aldrin, an insecticide used in soils to kill termites, grasshoppers, Western corn rootworm, and others, is also known to kill birds, fish, and humans. Humans are primarily exposed to aldrin through dairy products and animal meats.

2. Chlordane, an insecticide used to control termites and on a range of agricultural crops, is known to be lethal in various species of birds, including mallard ducks, bobwhite quail, and pink shrimp; it is a chemical that remains in the soil with a reported half-life of one year. Chlordane has been postulated to affect the human immune system and is classified as a possible human carcinogen. Chlordane air pollution is believed the primary route of humane exposure.

3. Dieldrin, a pesticide used to control termites, textile pests, insect-borne diseases and insects living in agricultural soils. In soil and insects, aldrin can be oxidized, resulting in rapid conversion to dieldrin. Dieldrin's half-life is approximately five years. Dieldrin is highly toxic to fish and other aquatic animals, particularly frogs, whose embryos can develop spinal deformities after exposure to low levels. Dieldrin has been linked to Parkinson's disease, breast cancer, and classified as immunotoxic, neurotoxic, with endocrine disrupting capacity. Dieldrin residues have been found in air, water, soil, fish, birds, and mammals. Human exposure to dieldrin primarily derives from food.

4. Endrin, an insecticide sprayed on the leaves of crops, and used to control rodents. Animals can metabolize endrin, so fatty tissue accumulation is not an issue, however the chemical has a long half-life in soil for up to 12 years. Endrin is highly toxic to aquatic animals and humans as a neurotoxin. Human exposure results primarily through food.

5. Heptachlor, a pesticide primarily used to kill soil insects and termites, along with cotton insects, grasshoppers, other crop pests, and malaria-carrying mosquitoes. Heptachlor, even at every low doses has been associated with the decline of several wild bird populations – Canada geese and American kestrels. In laboratory tests have shown high-dose heptachlor as lethal, with adverse behavioral changes and reduced reproductive success at low-doses, and is classified as a possible human carcinogen. Human exposure primarily results from food.

6. Hexachlorobenzene (HCB), was first introduced in 1945–1959 to treat seeds because it can kill fungi on food crops. HCB-treated seed grain consumption is associated with photosensitive skin lesions, colic, debilitation, and a metabolic disorder called porphyria turci-

ca, which can be lethal. Mothers who pass HCB to their infants through the placenta and breast milk had limited reproductive success including infant death. Human exposure is primarily from food.

7. Mirex, an insecticide used against ants and termites or as a flame retardant in plastics, rubber, and electrical goods. Mirex is one of the most stable and persistent pesticides, with a half-life of up to 10 years. Mirex is toxic to several plant, fish and crustacean species, with suggested carcinogenic capacity in humans. Humans are exposed primarily through animal meat, fish, and wild game.

8. Toxaphene, an insecticide used on cotton, cereal, grain, fruits, nuts, and vegetables, as well as for tick and mite control in livestock. Widespread toxaphene use in the US and chemical persistence, with a half-life of up to 12 years in soil, results in residual toxaphene in the environment. Toxaphene is highly toxic to fish, inducing dramatic weight loss and reduced egg viability. Human exposure primarily results from food. While human toxicity to direct toxaphene exposure is low, the compound is classified as a possible human carcinogen.

9. Polychlorinated biphenyls (PCBs), used as heat exchange fluids, in electrical transformers, and capacitors, and as additives in paint, carbonless copy paper, and plastics. Persistence varies with degree of halogenation, an estimated half-life of 10 years. PCBs are toxic to fish at high doses, and associated with spawning failure at low doses. Human exposure occurs through food, and is associated with reproductive failure and immune suppression. Immediate effects of PCB exposure include pigmentation of nails and mucous membranes and swelling of the eyelids, along with fatigue, nausea, and vomiting. Effects are transgenerational, as the chemical can persist in a mother's body for up to 7 years, resulting in developmental delays and behavioral problems in her children. Food contamination has led to large scale PCB exposure.

10. Dichlorodiphenyltrichloroethane (DDT) is probably the most infamous POP. It was widely used as insecticide during WWII to protect against malaria and typhus. After the war, DDT was used as an agricultural insecticide. In 1962, the American biologist Rachel Carson published Silent Spring, describing the impact of DDT spraying on the US environment and human health. DDT's persistence in the soil for up to 10–15 years after application has resulted in widespread and persistent DDT residues throughout the world including the arctic, even though it has been banned or severely restricted in most of the world. DDT is toxic to many organisms including birds where it is detrimental to reproduction due to eggshell thinning. DDT can be detected in foods from all over the world and food-borne DDT remains the greatest source of human exposure. Short-term acute effects of DDT on humans are limited, however long-term exposure has been associated with chronic health effects including increased risk of cancer and diabetes, reduced reproductive success, and neurological disease.

11. Dioxins are unintentional by-products of high-temperature processes, such as incomplete combustion and pesticide production. Dioxins are typically emitted from the burning of hospital waste, municipal waste, and hazardous waste, along with automobile emissions, peat, coal, and wood. Dioxins have been associated with several adverse effects in humans, including immune and enzyme disorders, chloracne, and are classified as a possible human

carcinogen. In laboratory studies of dioxin effects an increase in birth defects and still-births, and lethal exposure have been associated with the substances. Food, particularly from animals, is the principal source of human exposure to dioxins.

12. Polychlorinated dibenzofurans are by-products of high-temperature processes, such as incomplete combustion after waste incineration or in automobiles, pesticide production, and polychlorinated biphenyl production. Structurally similar to dioxins, the two compounds share toxic effects. Furans persist in the environment and classified as possible human carcinogens. Human exposure to furans primarily results from food, particularly animal products.

New POPs on the Stockholm Convention List

Since 2001, this list has been expanded to include some polycyclic aromatic hydrocarbons (PAHs), brominated flame retardants, and other compounds. Additions to the initial 2001 Stockholm Convention list are as following POPs:

- Chlordecone, a synthetic chlorinated organic compound,is primarily used as an agricultural pesticide, related to DDT and Mirex. Chlordecone is toxic to aquatic organisms, and classified as a possible human carcinogen. Many countries have banned chlordecone sale and use, or intend to phase out stockpiles and wastes.

- α-Hexachlorocyclohexane (α-HCH) and β-Hexachlorocyclohexane (β-HCH) are insecticides as well as by-products in the production of lindane. Large stockpiles of HCH isomers exist in the environment. α-HCH and β-HCH are highly persistent in the water of colder regions. α-HCH and β-HCH has been linked Parkinson's and Alzheimer's disease.

- Hexabromodiphenyl ether (hexaBDE) and heptabromodiphenyl ether (heptaBDE) are main components of commercial octabromodiphenyl ether (octaBDE). Commercial octaBDE is highly persistent in the environment, whose only degradation pathway is through debromination and the production of bromodiphenyl ethers, which can increase toxicity.

- Lindane (γ-hexachlorocyclohexane), a pesticide used as a broad spectrum insecticide for seed, soil, leaf, tree and wood treatment, and against ectoparasites in animals and humans (head lice and scabies). Lindane rapidly bioconcentrates. It is immunotoxic, neurotoxic, carcinogenic, linked to liver and kidney damage as well as adverse reproductive and developmental effects in laboratory animals and aquatic organisms. Production of lindane unintentionally produces two other POPs α-HCH and β-HCH.

- Pentachlorobenzene (PeCB), is a pesticide and unintentional byproduct. PeCB has also been used in PCB products, dyestuff carriers, as a fungicide, a flame retardant, and a chemical intermediate. PeCB is moderately toxic to humane, while highly toxic to aquatic organisms.

- Tetrabromodiphenyl ether (tetraBDE) and pentabromodiphenyl ether (pentaBDE) are industrial chemicals and the main components of commercial pentabromodiphenyl ether (pentaBDE). PentaBDE has been detected in humans in all regions of the world.

- Perfluorooctanesulfonic acid (PFOS) and its salts are used in the production of fluoropolymers. PFOS and related compounds are extremely persistent, bioaccumulating and bio-magnifying. The negative effects of trace levels of PFOS have not been established.

- Endosulfans are insecticides to control pests on crops such coffee, cotton, rice and sor-ghum and soybeans, tsetse flies, ectoparasites of cattle. They are used as a wood preserva-tive. Global use and manufacturing of endosulfan has been banned under the Stockholm convention in 2011, although many countries had previously banned or introduced phase-outs of the chemical when the ban was announced. Toxic to humans and aquatic and ter-restrial organisms, linked to congenital physical disorders, mental retardation, and death. Endosulfans' negative health effects are primarily liked to its endocrine disrupting capacity acting as an antiandrogen.

- Hexabromocyclododecane (HBCD) is a brominated flame retardant primarily used in ther-mal insulation in the building industry. HBCD is persistent, toxic and ecotoxic, with bioac-cumulative and long-range transport properties.

Additive and synergistic Effects

Evaluation of the effects of POPs on health is very challenging in the laboratory setting. For exam-ple, for organisms exposed to a mixture of POPs, the effects are assumed to be additive. Mixtures of POPs can in principle produce synergistic effects. With synergistic effects, the toxicity of each compound is enhanced (or depressed) by the presence of other compounds in the mixture. When put together, the effects can far exceed the approximated additive effects of the POP compound mixture.

Health Effects

POP exposure may cause developmental defects, chronic illnesses, and death. Some are carcino-gens per IARC, possibly including breast cancer. Many POPs are capable of endocrine disruption within the reproductive system, the central nervous system, or the immune system. People and animals are exposed to POPs mostly through their diet, occupationally, or while growing in the womb. For humans not exposed to POPs through accidental or occupational means, over 90% of exposure comes from animal product foods due to bioaccumulation in fat tissues and bioaccumu-late through the food chain. In general, POP serum levels increase with age and tend to be higher in females than males.

Studies have investigated the correlation between low level exposure of POPs and various diseases. In order to assess disease risk due to POPs in a particular location, government agencies may pro-duce a human health risk assessment which takes into account the pollutants' bioavailability and their dose-response relationships.

Endocrine Disruption

The majority of POPs are known to disrupt normal functioning of the endocrine system, for exam-ple all of the dirty dozen are endocrine disruptors. Low level exposure to POPs during critical de-velopmental periods of fetus, newborn and child can have a lasting effect throughout its lifespan. A 2002 study synthesizes data on endocrine disruption and health complications from exposure to POPs during critical developmental stages in an organism's lifespan. The study aimed to answer the question whether or not chronic, low level exposure to POPs can have a health impact on the endocrine system and development of organisms from different species. The study found that exposure of POPs during a critical developmental time frame can produce a permanent changes in the organisms path of development. Exposure of POPs during non-critical developmental time frames may not lead to detectable diseases and health complications later in their life. In wildlife, the critical development time frames are in utero, in ovo, and during reproductive periods. In humans, the critical development timeframe is during fetal development.

Reproductive System

The same study in 2002 with evidence of a link from POPs to endocrine disruption also linked low dose exposure of POPs to reproductive health effects. The study stated that POP exposure can lead to negative health effects especially in the male reproductive system, such as decreased sperm quality and quantity, altered sex ratio and early puberty onset. For females exposed to POPs, altered reproductive tissues and pregnancy outcomes as well as endometriosis have been reported.

Exposure During Pregnancy

POP exposure during pregnancy is of particular concern to the developing fetus.

Transport Across the Placenta

A study about the transfer of POPs (14 organochlorine pesticides, 7 polychlorinated biphenyls and 14 polybrominated diphenyl ethers (PBDEs)) from Spanish mothers to their unborn fetus found that POP concentrations in serum from the mother were higher than from the umbilical cord and 50 placentas. Because transfer of the POPs from mother to fetus did not correspond with passive lipid-associated diffusion, authors suggested that POPs are actively transported across the placenta.

Gestational Weight Gain and Newborn Head Circumference

A Greek study from 2014 investigated the link between maternal weight gain during pregnancy, their PCB-exposure level and PCB level in their newborn infants, their birth weight, gestational age, and head circumference. The birth weight and head circumference of the infants was the lower, the higher POP levels during prenatal development had been, but only if mothers had either excessive or inadequate weight gain during pregnancy. No correlation between POP exposure and gestational age was found. A 2013 case-control study conducted 2009 in Indian mothers and their offspring showed prenatal exposure of two types of organochlorine pesticides (HCH, DDT and DDE) impaired the growth of the fetus, reduced the birth weight, length, head circumference and chest circumference.

Cardiovascular Disease and Cancer

POPs are lipophilic environmental toxins. They are often found in lipoproteins of organisms. A study published in 2014 found an association between the concentration of POPs in lipoproteins and the occurrence of cardiovascular disease and various cancers in human beings. The higher the concentration of POPs found in lipoproteins, the higher the occurrence of cardiovascular disease and cancer. Highly chlorinated polychlorinated biphenyls are specifically found in high concentrations in lipoproteins. Cardiovascular disease is shown to be more associated with higher concentrations of POPs in high density lipoproteins and cancer is shown to be more associated with higher concentrations of POPs in low density lipoproteins and very low density lipoproteins.

Obesity

There have been many recent studies assessing the connection between serum POP levels in individuals and instances of obesity. A study released in 2011 found correlations between different POPs and obesity occurrence in individuals tested. The statistically significant findings from the study show that there is actually a negative correlation between various PCB congener serum levels and obesity in individuals tested. The study also showed a positive correlation between beta-hexachlorocyclohexane and various dioxin serum levels and obesity in individuals tested. Obesity was determined using the Body Mass Index (BMI). One proposed explanation in the study is that PCBs are very lipophilic, therefore they are easily stored and captured in the fat deposits in human beings. Obese individuals have higher amounts of fat deposits in their body, and thus more PCBs could be captured in the fat deposits leading to less PCBs circulating in blood serum. The study provides evidence demonstrating that the correlation between POP serum levels and obesity occurrence is more complicated than previously expected. The same study also noted a strong positive correlation between serum POP levels and age in all individuals in the experiment.

Diabetes

A study published in 2006 revealed a positive correlation between POP serum levels and type II diabetes in individuals, after other variables, such as age, sex, race, and socioeconomic status were adjusted for. The correlation proved stronger in younger, Mexican American, and obese individuals. Individuals exposed to low doses of POPs throughout their lifetime had a higher chance for developing diabetes than individuals exposed to high concentrations of POPs for a short amount of time.

POPs in Urban Areas and Indoor Environments

Traditionally it was thought that human exposure to POPs occurred primarily through food, however indoor pollution patterns that characterize certain POPs have challenged this notion. Recent studies of indoor dust and air have implicated indoor environments as a major sources for human exposure via inhalation and ingestion. Furthermore, significant indoor POP pollution must be a major route of human POP exposure, considering the modern trend in spending larger proportions of life indoor. Several studies have shown that indoor (air and dust) POP levels to exceed outdoor (air and soil) POP concentrations.

Control and Removal of POPs in the Environment

Current studies aimed at minimizing POPs in the environment are investigating their behavior in photo catalytic oxidation reactions. POPs that are found in humans and in aquatic environments the most are the main subjects of these experiments. Aromatic and aliphatic degradation products have been identified in these reactions. Photochemical degradation is negligible compared to photocatalytic degradation. However, proper removal techniques of POPs from the environment are still unclear, due to fear that more toxic byproducts may result from uninvestigated degradation techniques. Current efforts are more focused on banning the use and production of POPs worldwide rather than removal of POPs.

Health Effects of Pesticides

Health effects of pesticides may be acute or delayed in those who are exposed. A 2007 systematic review found that "most studies on non-Hodgkin lymphoma and leukemia showed positive associations with pesticide exposure" and thus concluded that cosmetic use of pesticides should be decreased. Strong evidence also exists for other negative outcomes from pesticide exposure including neurological problems, birth defects, fetal death, and neurodevelopmental disorder.

According to The Stockholm Convention on Persistent Organic Pollutants, 9 of the 12 most dangerous and persistent chemicals are pesticides.

Acute Effects

Acute health problems may occur in workers that handle pesticides, such as abdominal pain, dizziness, headaches, nausea, vomiting, as well as skin and eye problems. In China, an estimated half million people are poisoned by pesticides each year, 500 of whom die. Pyrethrins, insecticides commonly used in common bug killers, can cause a potentially deadly condition if breathed in.

Long-term Effects

Cancer

Many studies have examined the effects of pesticide exposure on the risk of cancer. Associations have been found with: leukemia, lymphoma, brain, kidney, breast, prostate, pancreas, liver, lung, and skin cancers. This increased risk occurs with both residential and occupational exposures. Increased rates of cancer have been found among farm workers who apply these chemicals. A mother's occupational exposure to pesticides during pregnancy is associated with an increases in her child's risk of leukemia, Wilms' tumor, and brain cancer. Exposure to insecticides within the home and herbicides outside is associated with blood cancers in children.

Neurological

Evidence links pesticide exposure to worsened neurological outcomes. The risk of developing Parkinson's disease is 70% greater in those exposed to even low levels of pesticides. People with Parkinson's were 61% more likely to report direct pesticide application than were healthy relatives. Both insecticides and herbicides significantly increased the risk of Parkinson's disease. There are also concerns that long-term exposures may increase the risk of dementia.

The United States Environmental Protection Agency finished a 10-year review of the organophosphate pesticides following the 1996 Food Quality Protection Act, but did little to account for developmental neurotoxic effects, drawing strong criticism from within the agency and from outside researchers. Comparable studies have not been done with newer pesticides that are replacing organophosphates.

Reproductive Effects

Strong evidence links pesticide exposure to birth defects, fetal death and altered fetal growth. In the United States, increase in birth defects is associated with conceiving in the same period of the year when agrochemicals are in elevated concentrations in surface water. Agent Orange, a 50:50 mixture of 2,4,5-T and 2,4-D, has been associated with bad health and genetic effects in Malaya and Vietnam. It was also found that offspring that were at some point exposed to pesticides had a low birth weight and had developmental defects.

Fertility

A number of pesticides including dibromochlorophane and 2,4-D has been associated with impaired fertility in males. Pesticide exposure resulted in reduced fertility in males, genetic alterations in sperm, a reduced number of sperm, damage to germinal epithelium and altered hormone function.

Other

Some studies have found increased risks of dermatitis in those exposed.

Additionally, studies have indicated that pesticide exposure is associated with long-term health problems such as respiratory problems, including asthma, memory disorders and depression. Summaries of peer-reviewed research have examined the link between pesticide exposure and neurologic outcomes and cancer, perhaps the two most significant things resulting in organophosphate-exposed workers.

According to researchers from the National Institutes of Health (NIH), licensed pesticide applicators who used chlorinated pesticides on more than 100 days in their lifetime were at greater risk of diabetes. One study found that associations between specific pesticides and incident diabetes ranged from a 20 percent to a 200 percent increase in risk. New cases of diabetes were reported by 3.4 percent of those in the lowest pesticide use category compared with 4.6 percent of those in the highest category. Risks were greater when users of specific pesticides were compared with applicators who never applied that chemical.

Route of Exposure

People can be exposed to pesticides by a number of different routes including: occupation, in the home, at school and in their food.

There are concerns that pesticides used to control pests on food crops are dangerous to people who consume those foods. These concerns are one reason for the organic food movement. Many food crops, including fruits and vegetables, contain pesticide residues after being washed or peeled. Chemicals that are no longer used but that are resistant to breakdown for long periods may remain in soil and water and thus in food.

The United Nations Codex Alimentarius Commission has recommended international standards for maximum residue limits (MRLs), for individual pesticides in food.

In the EU, MRLs are set by DG-SANCO.

In the United States, levels of residues that remain on foods are limited to tolerance levels that are established by the U.S. Environmental Protection Agency and are considered safe. The EPA sets the tolerances based on the toxicity of the pesticide and its breakdown products, the amount and frequency of pesticide application, and how much of the pesticide (i.e., the residue) remains in or on food by the time it is marketed and prepared. Tolerance levels are obtained using scientific risk assessments that pesticide manufacturers are required to produce by conducting toxicological studies, exposure modeling and residue studies before a particular pesticide can be registered, however, the effects are tested for single pesticides, and there is little information on possible synergistic effects of exposure to multiple pesticide traces in the air, food and water.

Strawberries and tomatoes are the two crops with the most intensive use of soil fumigants. They are particularly vulnerable to several type of diseases, insects, mites, and parasitic worms. In 2003, in California alone, 3.7 million pounds (1,700 metric tons) of metham sodium were used on tomatoes. In recent years other farmers have demonstrated that it is possible to produce strawberries and tomatoes without the use of harmful chemicals and in a cost-effective way.

Exposure routes other than consuming food that contains residues, in particular pesticide drift, are potentially significant to the general public.

Some pesticides can remain in the environment for prolonged periods of time. For example, most people in the United States still have detectable levels of DDT in their bodies even though it was banned in the US in 1972.

Prevention

Pesticides exposure cannot be studied in placebo controlled trials as this would be unethical. A definitive cause effect relationship therefore cannot be established. Consistent evidence can and has been gathered through other study designs. The precautionary principle is thus frequently used in environmental law such that absolute proof is not required before efforts to decrease exposure to potential toxins are enacted.

The American Medical Association recommend limiting exposure to pesticides. They came to this conclusion due to the fact that surveillance systems currently in place are inadequate to determine problems related to exposure. The utility of applicator certification and public notification programs are also of unknown value in their ability to prevent adverse outcomes.

Epidemiology

The World Health Organization and the UN Environment Programme estimate that each year, 3 million workers in agriculture in the developing world experience severe poisoning from pesticides, about 18,000 of whom die. According to one study, as many as 25 million workers in developing countries may suffer mild pesticide poisoning yearly. Detectable levels of 50 different pesticides were found in the blood of a representative sample of the U.S. population.

Society and Culture

Concerns regarding conflict of interests regarding the research base have been raised. A number of researchers involved with pesticides have been found to have undisclosed ties to industry including: Richard Doll or the Imperial Cancer Research Fund in England and Hans-Olov Adami of the Karolinska Institute in Sweden.

Other Animals

A number of pesticides including clothianidin, dinotefuran, imidacloprid are toxic to bees. Exposure to pesticides may be one of the contributory factors to colony collapse disorder. A study in North Carolina indicated that more than 30 percent of the quail tested were made sick by one aerial insecticide application. Once sick, wild birds may neglect their young, abandon their nests, and become more susceptible to predators or disease.

Pesticide Poisoning

A pesticide poisoning occurs when chemicals intended to control a pest affect non-target organisms such as humans, wildlife, or bees. There are three types of pesticide poisoning. The first of the three is a single and short-term very high level of exposure which can be experienced by individuals who commit suicide, as well as pesticide formulators. The second type of poisoning is long-term high-level exposure, which can occur in pesticide formulators and manufacturers. The third type of poisoning is a long-term low-level exposure, which individuals are exposed to from sources such as pesticide residues in food as well as contact with pesticide residues in the air, water, soil, sediment, food materials, plants and animals.

In developing countries, such as Sri Lanka, pesticide poisonings from short-term very high level of exposure (acute poisoning) is the most worrisome type of poisoning. However, in developed countries, such as Canada, it is the complete opposite: acute pesticide poisoning is controlled, thus making the main issue long-term low-level exposure of pesticides.

Cause

The most common exposure scenarios for pesticide-poisoning cases are accidental or suicidal poisonings, occupational exposure, by-stander exposure to off-target drift, and the general public who are exposed through environmental contamination.

Accidental and Suicidal

Self-poisoning with agricultural pesticides represents a major hidden public health problem accounting for approximately one-third of all suicides worldwide. It is one of the most common forms of self-injury in the Global South. The World Health Organization estimates that 300,000 people die from self-harm each year in the Asia-Pacific region alone. Most cases of intentional pesticide poisoning appear to be impulsive acts undertaken during stressful events, and the availability of pesticides strongly influences the incidence of self poisoning. Pesticides are the agents most frequently used by farmers and students in India to commit suicide.

Occupational

Pesticide poisoning is an important occupational health issue because pesticides are used in a large number of industries, which puts many different categories of workers at risk. Extensive use puts agricultural workers in particular at increased risk for pesticide illnesses. Workers in other industries are at risk for exposure as well. For example, commercial availability of pesticides in stores puts retail workers at risk for exposure and illness when they handle pesticide products. The ubiquity of pesticides puts emergency responders such as fire-fighters and police officers at risk, because they are often the first responders to emergency events and may be unaware of the presence of a poisoning hazard. The process of aircraft disinsection, in which pesticides are used on inbound international flights for insect and disease control, can also make flight attendants sick.

Different job functions can lead to different levels of exposure. Most occupational exposures are caused by absorption through exposed skin such as the face, hands, forearms, neck, and chest. This exposure is sometimes enhanced by inhalation in settings including spraying operations in greenhouses and other closed environments, tractor cabs, and the operation of rotary fan mist sprayers.

Residential

When thinking of pesticide poisoning, one does not take into consideration the contribution that is made of their own household. The majority of households in Canada use pesticides while taking part in activities such as gardening. In Canada 96 percent of households report having a lawn or a garden. 56 percent of the households who have a lawn or a garden utilize fertilizer or pesticide. This form of pesticide use may contribute to the third type of poisoning, which is caused by long-term low-level exposure. As mentioned before, long-term low-level exposure affects individuals from sources such as pesticide residues in food as well as contact with pesticide residues in the air, water, soil, sediment, food materials, plants and animals.

Pathophysiology

Organochlorines

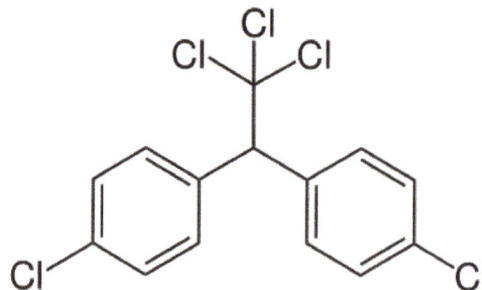

DDT, an organochlorine

The organochlorine pesticides, like DDT, aldrin, and dieldrin are extremely persistent and accumulate in fatty tissue. Through the process of bioaccumulation (lower amounts in the environment get magnified sequentially up the food chain), large amounts of organochlorines can accumulate in top species like humans. There is substantial evidence to suggest that DDT, and its metabolite DDE, act as endocrine disruptors, interfering with hormonal function of estrogen, testosterone, and other steroid hormones.

Anticholinesterase Compounds

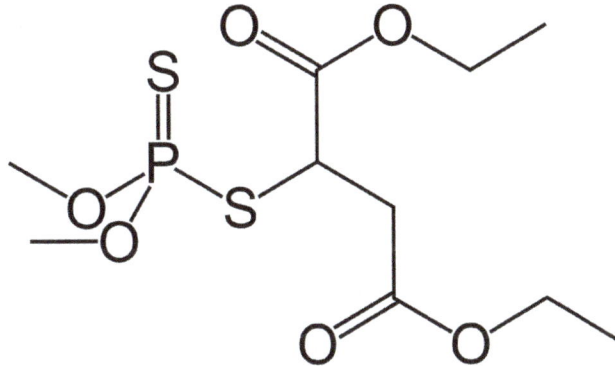

Malathion, an organophosphate anticholinesterase

Certain organophosphates have long been known to cause a delayed-onset toxicity to nerve cells, which is often irreversible. Several studies have shown persistent deficits in cognitive function in workers chronically exposed to pesticides. Newer evidence suggests that these pesticides may cause developmental neurotoxicity at much lower doses and without depression of plasma cholinesterase levels.

Diagnosis

Most pesticide-related illnesses have signs and symptoms that are similar to common medical conditions, so a complete and detailed environmental and occupational history is essential for correctly diagnosing a pesticide poisoning. A few additional screening questions about the patient's work and home environment, in addition to a typical health questionnaire, can indicate whether there was a potential pesticide poisoning.

If one is regularly using carbamate and organophosphate pesticides, it is important to obtain a baseline cholinesterase test. Cholinesterase is an important enzyme of the nervous system, and these chemical groups kill pests and potentially injure or kill humans by inhibiting cholinesterase. If one has had a baseline test and later suspects a poisoning, one can identify the extent of the problem by comparison of the current cholinesterase level with the baseline level.

Prevention

Accidental poisonings can be avoided by proper labeling and storage of containers. When handling or applying pesticides, exposure can be significantly reduced by protecting certain parts of the body where the skin shows increased absorption, such as the scrotal region, underarms, face, scalp, and hands. Using chemical-resistant gloves has been shown to reduce contamination by 33–86%.

Further methods in order to aid prevention of acute pesticide poisoning, concerning both accidental death and suicides, there could be a method for national governments to control accessibility. The pesticides most toxic to humans if restricted has the possibility to reduce deaths. There could also be designated locations in rural living areas and cities used to safely store toxic pesticides in order to gain control over usage.

Treatment

Specific treatments for acute pesticide poisoning are often dependent on the pesticide or class of pesticide responsible for the poisoning. However, there are basic management techniques that are applicable to most acute poisonings, including skin decontamination, airway protection, gastrointestinal decontamination, and seizure treatment.

Decontamination of the skin is performed while other life-saving measures are taking place. Clothing is removed, the patient is showered with soap and water, and the hair is shampooed to remove chemicals from the skin and hair. The eyes are flushed with water for 10–15 minutes. The patient is intubated and oxygen administered, if necessary. In more severe cases, pulmonary ventilation must sometimes be supported mechanically. Seizures are typically managed with lorazepam, phenytoin and phenobarbitol, or diazepam (particularly for organochlorine poisonings).

Gastric lavage is not recommended to be used routinely in pesticide poisoning management, as clinical benefit has not been confirmed in controlled studies; it is indicated only when the patient has ingested a potentially life-threatening amount of poison and presents within 60 minutes of ingestion. An orogastric tube is inserted and the stomach is flushed with saline to try to remove the poison. If the patient is neurologically impaired, a cuffed endotracheal tube inserted beforehand for airway protection. Studies of poison recovery at 60 minutes have shown recovery of 8%–32%. However, there is also evidence that lavage may flush the material into the small intestine, increasing absorption. Lavage is contra-indicated in cases of hydrocarbon ingestion.

Activated charcoal is sometimes administered as it has been shown to be successful with some pesticides. Studies have shown that it can reduce the amount absorbed if given within 60 minutes, though there is not enough data to determine if it is effective if time from ingestion is prolonged. Syrup of ipecac is no longer recommended for most pesticide poisonings.

Urinary alkalinisation has been used in acute poisonings from chlorophenoxy herbicides (such as 2,4-D, MCPA, 2,4,5-T and mecoprop); however, evidence to support its use is poor.

Epidemiology

Acute pesticide poisoning is a large-scale problem, especially in developing countries.

"Most estimates concerning the extent of acute pesticide poisoning have been based on data from hospital admissions which would include only the more serious cases. The latest estimate by a WHO task group indicates that there may be 1 million serious unintentional poisonings each year and in addition 2 million people hospitalized for suicide attempts with pesticides. This necessarily reflects only a fraction of the real problem. On the basis of a survey of self-reported minor poisoning carried out in the Asian region, it is estimated that there could be as many as 25 million agricultural workers in the developing world suffering an episode of poisoning each year." In Canada in 2007 more than 6000 cases of acute pesticide poisoning occurred.

Estimating the numbers of chronic poisonings worldwide is more difficult.

Society and Culture

Rachel Carson's *Silent Spring* brought about the first major wave of public concern over the chronic effects of pesticides.

In other Animals

An obvious side effect of using a chemical meant to kill is that one is likely to kill more than just the desired organism. Contact with a sprayed plant or "weed" can have an effect upon local wildlife, most notably insects. A cause for concern is how pests, the reason for pesticide use, are building up a resistance. Phytophagous insects are able to build up this resistance because they are easily capable of evolutionary diversification and adaptation. The problem this presents is that in order to obtain the same desired effect of the pesticides they have to be made increasingly stronger as time goes on. Repercussions of the use of stronger pesticides on vegetation has a negative result on the surrounding environment, but also would contribute to consumers' long-term low-level exposure.

Toxicity Class

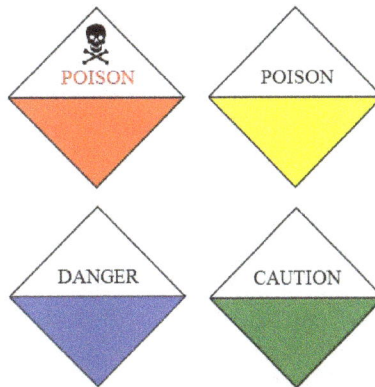

Indian toxicity label system

Toxicity symbol for European Toxicity Class I and II

Toxicity class refers to a classification system for pesticides that has been created by a national or international government-related or -sponsored organization. It addresses the acute toxicity of agents such as soil fumigants, fungicides, herbicides, insecticides, miticides, molluscicides, nematicides, or rodenticides.

General Considerations

Assignment to a toxicity class is based typically on results of acute toxicity studies such as the determination of LD_{50} values in animal experiments, notably rodents, via oral, inhaled, or external application. The experimental design measures the acute death rate of an agent. The toxicity class generally does not address issues of other potential harm of the agent, such as bioaccumulation, issues of carcinogenicity, teratogenicity, mutagenic effects, or the impact on reproduction.

Regulating agencies may require that packaging of the agent be labeled with a **signal word**, a specific warning label to indicate the level of toxicity.

Toxicity Class by Jurisdiction

World Health Organization

The World Health Organization (WHO) names four toxicity classes:

- Class I – a: extremely hazardous
- Class I – b: highly hazardous
- Class II: moderately hazardous
- Class III: slightly hazardous

The system is based on LD50 determination in rats, thus an oral solid agent with an LD50 at 5 mg or less/kg bodyweight is Class I-a, at 5–50 mg/kg Class I-b, at 50–2000 mg/kg Class II, and at more than 2000 mg/kg Class III. Values may differ for liquid oral agents and dermal agents.

European Union

There are eight toxicity classes in the European Union's classification system, which is regulated by Directive 67/548/EEC:

- Class I: very toxic
- Class II: toxic
- Class III: harmful
- Class IV : corrosive
- Class V : irritant
- Class VI : sensitizing
- Class VII : carcinogenic
- Class VIII : mutagenic

Very toxic and toxic substances are marked by the European toxicity symbol.

India

The Indian standardized system of toxicity labels for pesticides uses a 4-color system (red, yellow, blue, green) to plainly label containers with the toxicity class of the contents.

United States

The United States Environmental Protection Agency (EPA) uses four toxicity classes. Classes I to III are required to carry a signal word on the label. Pesticides are regulated in the United States primarily by the Federal Insecticide, Fungicide, and Rodenticide Act (FIFRA).

Toxicity Class I

- most toxic;
- requires signal word: "Danger-Poison", with skull and crossbones symbol, possibly followed by:

  "Fatal if swallowed", "Poisonous if inhaled", "Extremely hazardous by skin contact--rapidly absorbed through skin", or "Corrosive--causes eye damage and severe skin burns"

Class I materials are estimated to be fatal to an adult human at a dose of less than 5 grams (less than a teaspoon).

Toxicity Class II

- moderately toxic
- Signal word: "Warning", possibly followed by:

  "Harmful or fatal if swallowed", "Harmful or fatal if absorbed through the skin", "Harmful or fatal if inhaled", or "Causes skin and eye irritation"

Class II materials are estimated to be fatal to an adult human at a dose of 5 to 30 grams.

Toxicity Class III

- slightly toxic
- Signal word: Caution, possibly followed by:

  "Harmful if swallowed", "May be harmful if absorbed through the skin", "May be harmful if inhaled", or "May irritate eyes, nose, throat, and skin"

Class III materials are estimated to be fatal to an adult human at some dose in excess of 30 grams.

Toxicity Class IV

- practically nontoxic
- no Signal Word required since 2002

General Versus Restricted Use

Furthermore, the EPA classifies pesticides into those anybody can apply (*General Use Pesticides*), and those that must be applied by or under the supervision of a certified individual. Application of *Restricted use pesticides* requires that a record of the application be kept.

Pesticide Toxicity to Bees

Pesticides vary in their effects on bees. Contact pesticides are usually sprayed on plants and can kill bees when they crawl over sprayed surfaces of plants or other areas around it. Systemic pesticides, on the other hand, are usually incorporated into the soil or onto seeds and move up into the stem, leaves, nectar, and pollen of plants.

Dead bees in a French beekeeping farm. Credit: Raymond Roig

Of contact pesticides, dust and wettable powder pesticides tend to be more hazardous to bees than solutions or emulsifiable concentrates. When a bee comes in contact with pesticides while foraging, the bee may die immediately without returning to the hive. In this case, the queen bee, brood, and nurse bees are not contaminated and the colony survives. Alternatively, the bee may come into contact with an insecticide and transport it back to the colony in contaminated pollen or nectar or on its body, potentially causing widespread colony death.

Actual damage to bee populations is a function of toxicity and exposure of the compound, in combination with the mode of application. A systemic pesticide, which is incorporated into the soil or coated on seeds, may kill soil-dwelling insects, such as grubs or mole crickets as well as other insects, including bees, that are exposed to the leaves, fruits, pollen, and nectar of the treated plants.

Pesticides are linked to Colony Collapse Disorder and are now considered a main cause, and the toxic effects of Neonicotinoids on bees are confirmed. Currently, many studies are being conducted to further understand the toxic effects of pesticides on bees. Agencies such as the EPA and EFSA are making action plans to protect bee health in response to calls from scientists and the public to ban or limit the use of the pesticides with confirmed toxicity.

Classification

Insecticide toxicity is generally measured using acute contact toxicity values LD_{50} – the exposure level that causes 50% of the population exposed to die. Toxicity thresholds are generally set at

- highly toxic (acute LD50 < 2µg/bee)

- moderately toxic (acute LD50 2 - 10.99µg/bee)

- slightly toxic (acute LD50 11 - 100µg/bee)

- nontoxic (acute LD50 > 100µg/bee) to adult bees.

Pesticide Toxicity

Acute Toxicity

The acute toxicity of pesticides on bees, which could be by contact or ingestion, is usually quantified by LD_{50}. Acute toxicity of pesticides causes a range of effects on bees, which can include agitation, vomiting, wing paralysis, arching of the abdomen similar to sting reflex, and uncoordinated movement. Some pesticides, including Neonicotinoids, are more toxic to bees and cause acute symptoms with lower doses compared to older classes of insecticides. Acute toxicity may depend on the mode of exposure, for instance, many pesticides cause toxic effects by contact while Neonicotinoids are more toxic when consumed orally. The acute toxicity, although more lethal, is less common than sub-lethal toxicity or cumulative effects.

Sub-lethal Toxicity

Field exposure of bees to pesticides, especially with relation to neonicotinoids, is most commonly sub-lethal. Sub-lethal effects to honey bees are of major concern and include behavioral disruptions such as disorientation, reduced foraging, impaired memory and learning, and a shift in communication behaviors. Additional sub-lethal effects may include compromised immunity of bees and delayed development.

Cumulative and Chronic Effects

Neonicotinoids are especially likely to cause cumulative effects on bees due to their mechanism of function as this pesticide group works by binding to nicotinic acetylcholine receptors in the brains of the insects, and such receptors are particularly abundant in bees. Over-accumulation of acetylcholine results in paralysis and death.

Colony Collapse Disorder

Colony collapse disorder (CCD) is a syndrome that is characterized by the sudden loss of adult bees from the hive. Many possible explanations for CCD have been proposed, but no one primary cause has been found. The US Department of Agriculture (USDA) has indicated in a report to Congress that a combination of factors may be causing CCD, including pesticides, pathogens, and parasites, all of which have been found at high levels in affected bee hives.

Signs promote awareness of Colony Collapse Disorder and the importance of bees.

Colony Collapse Disorder has more implication than the extinction of some bee species; the disappearance of honeybees can cause catastrophic health and financial impacts. One mouthful in three of the food we eat may depend directly or indirectly on pollination by honeybee. Honeybee pollination has an estimated value of more than $14 billion annually to the United States agriculture. Honeybees are required for pollinating many crops, which range from nuts to vegetables and fruits, that are necessary for human and animal diet.

The EPA updated their guidance for assessing pesticide risks to honeybees in 2014. For the EPA, when certain pesticide use patterns or triggers are met, current test requirements include the honey bee acute contact toxicity test, the honey bee toxicity of residues on foliage test, and field testing for pollinators. EPA guidelines have not been developed for chronic or acute oral toxicity to adult or larval honey bees. On the other hand, the PMRA (Pest Management Regulatory Agency) requires both acute oral and contact honey bee adult toxicity studies when there is potential for exposure for insect pollinators. Primary measurement endpoint derived from the acute oral and acute contact toxicity studies is the median lethal dose for 50% of the organisms tested (i.e., LD_{50}), and if any biological effects and abnormal responses appear, including sub-lethal effects, other than the mortality, it should be reported.

The EPA's testing requirements do not account for sub-lethal effects to bees or effects on brood or larvae. Their testing requirements are also not designed to determine effects in bees from exposure to systemic pesticides. With colony collapse disorder, whole hive tests in the field are needed in order to determine the effects of a pesticide on bee colonies. To date, there are very few scientifically valid whole hive studies that can be used to determine the effects of pesticides on bee colonies because the interpretation of such whole-colony effects studies is very complex and relies on comprehensive considerations of whether adverse effects are likely to occur at the colony level.

A March 2012 study conducted in Europe, in which minuscule electronic localization devices were fixed on bees, has shown that, even with very low levels of pesticide in the bee's diet, a high proportion of bees (more than one third) suffers from orientation disorder and is unable to come back to the hive. The pesticide concentration was order of magnitudes smaller than the lethal dose used in the pesticide's current use. The pesticide under study, brand-named "Cruiser" in Europe (thiamethoxam, a neonicotinoid insecticide), although allowed in France by annually renewed exceptional authorization, could be banned in the coming years by the European Commission.

April 2013 the EU decided to restrict thiamethoxam, clothianidin, and imidacloprid.

Bee Kill Rate Per Hive

The kill rate of bees in a single bee hive can be classified as:

> < 100 bees per day - normal die off rate

> 200-400 bees per day - low kill

> 500-900 bees per day - moderate kill

> > 1000 bees per day - high kill

Pesticides Formulations

Pesticides come in different formulations:

- Dusts (D)

- Wettable powders (WP)

- Soluble powders (SP)

- Emulsifiable concentrates (EC)

- Solutions (LS)

- Granulars (G)

Pesticides

Common name (ISO)	Examples of Brand names	Pesticide Class	length of residual toxicity	Comments	Bee toxicity
Sulfoxaflor		Sulfoximine			
Aldicarb	Temik	Carbamate		apply 4 weeks before bloom	Relatively nontoxic
Carbaryl	Sevin, (b) Sevin XLR	Carbamate	High risk to bees foraging even 10 hours after spraying; 3 – 7 days (b) 8 hours @ 1.5 lb/acre (1681 g/Ha) or less.	Bees poisoned with carbaryl can take 2–3 days to die, appearing inactive as if cold. Sevin should never be sprayed on flowering crops, especially if bees are active and the crop requires pollination. Less toxic formulations exist.	highly toxic

Carbofuran	Furadan	Carbamate	7 – 14 days	U.S. Environmental Protection Agency ban on use on crops grown for human consumption (2009) carbofuran (banned in granular form)	highly toxic
Methomyl	Lannate, Nudrin	Carbamate	2 hours	Should never be sprayed on flowering crops especially if bees are active and the crop requires pollination.	highly toxic
Methiocarb	Mesurol	Carbamate			highly toxic
Mexacarbate	Zectran	Carbamate			highly toxic
Pirimicarb	Pirimor, Aphox	Carbamate			Relatively nontoxic
Propoxur	Baygon	Carbamate		Propoxur is highly toxic to honey bees. The LD50 for bees is greater than one ug/honey bee.	highly toxic
Acephate	Orthene	Organophosphate	3 days	Acephate is a broad-spectrum insecticide and is highly toxic to bees and other beneficial insects.	Moderately toxic
Azinphos-methyl	Guthion, Methyl-Guthion	Organophosphate	2.5 days	banned in the European Union since 2006.	highly toxic
Chlorpyrifos	Dursban, Lorsban	Organophosphate		banned in the US for home and garden use Should never be sprayed on flowering crops especially if bees are active and the crop requires pollination.	highly toxic
Coumaphos	Checkmite	Organophosphate		This is an insecticide that is used inside the beehive to combat varroa mites and small hive beetles, which are parasites of the honey bee. Overdoses can lead to bee poisoning.	Relatively nontoxic
Demeton	Systox	Organophosphate	<2 hours		highly toxic
Demeton-S-methyl	Meta-systox	Organophosphate			Moderately toxic

Diazinon	Spectracide	Organophosphate		Sale of diazinon for residential use was discontinued in the U.S. in 2004. Should never be sprayed on flowering crops especially if bees are active and the crop requires pollination.	highly toxic
Dicrotophos	Bidrin	Organophosphate		Dicrotophos toxicity duration is about one week.	highly toxic
Dichlorvos	DDVP, Vapona	Organophosphate			highly toxic
Dimethoate	Cygon, De-Fend	Organophosphate	3 days	Should never be sprayed on flowering crops especially if bees are active and the crop requires pollination.	highly toxic
Fenthion	Entex, Baytex, Baycid, Dalf, DMPT, Mercaptophos, Prentox, Fenthion 4E, Queletox, Lebaycid	Organophosphate		Should never be sprayed on flowering crops especially if bees are active and the crop requires pollination.	highly toxic
Fenitrothion	Sumithion	Organophosphate			highly toxic
Fensulfothion	Dasanit	Organophosphate			highly toxic
Fonofos	Dyfonate EC	Organophosphate	3 hours	List of Schedule 2 substances (CWC)	highly toxic
Malathion	Malathion USB, ~ EC, Cythion, maldison, mercaptothion	Organophosphate	>8 fl oz/ acre (58 L/ km^2) $\Rightarrow$ 5.5 days	Malathion is highly toxic to bees and other beneficial insects, some fish, and other aquatic life. Malathion is moderately toxic to other fish and birds, and is considered low in toxicity to mammals.	highly toxic
Methamidophos	Monitor, Tameron	Organophosphate		Should never be sprayed on flowering crops especially if bees are active and the crop requires pollination.	highly toxic
Methidathion	Supracide	Organophosphate		Should never be sprayed on flowering crops especially if bees are active and the crop requires pollination.	highly toxic

Methyl parathion	Parathion, Penncap-M	Organophosphate	5–8 days	It is classified as a UNEP Persistent Organic Pollutant and WHO Toxicity Class, "Ia, Extremely Hazardous".	highly toxic
Mevinphos	Phosdrin	Organophosphate			highly toxic
Monocrotophos	Azodrin	Organophosphate		Should never be sprayed on flowering crops especially if bees are active and the crop requires pollination.	highly toxic
Naled	Dibrom	Organophosphate	16 hours		highly toxic
Omethoate		Organophosphate		Should never be sprayed on flowering crops especially if bees are active and the crop requires pollination.	highly toxic
Oxydemeton-methyl	Metasystox-R	Organophosphate	<2 hours		highly toxic
Phorate	Thimet EC	Organophosphate	5 hours		highly toxic
Phosmet	Imidan	Organophosphate		Phosmet is very toxic to honeybees.	highly toxic
Phosphamidon	Dimecron	Organophosphate			highly toxic
Pyrazophos	Afugan	Organophosphate	fungicide		highly toxic
Tetrachlorvin-phos	Rabon, Stirofos, Gardona, Gardcide	Organophosphate			highly toxic
Trichlorfon, Metrifonate	Dylox, Dipterex	Organophosphate		3 – 6 hours	Relatively nontoxic
Permethrin	Ambush, Pounce	Synthetic pyrethroid	1 – 2 days	safened by repellency under arid conditions. Permethrin is also the active ingredient in insecticides used against the Small hive beetle, which is a parasite of the beehive in the temperate climate regions.	highly toxic
Cypermethrin	Ammo, Raid	Synthetic pyrethroid	Less than 2 hours	Cypermethrin is found in many household ant and cockroach killers, including Raid and ant chalk.	highly toxic

Fenvalerate	Asana, Pydrin	Synthetic pyrethroid	1 day	safened by repellency under arid conditions	highly toxic
Resmethrin	Chrysron, Crossfire, Pynosect, Raid Flying Insect Killer, Scourge, Sun-Bugger #4, SPB-1382, Synthrin, Syntox, Vectrin, Whitmire PT-110	Synthetic pyrethroid		Resmethrin is highly toxic to bees, with an LD50 of 0.063 ug/bee.	highly toxic
Methoxychlor	DMDT, Marlate	Chlorinated cyclodiene	2 hours	available as a General Use Pesticide	highly toxic
Endosulfan	Thiodan	Chlorinated cyclodiene	8 hours	banned in European Union (2007?), New Zealand (2009)	moderately toxic
Clothianidin	Poncho	Neonicotinoid		Banned in Germany In June 2008, the Federal Ministry of Food, Agriculture and Consumer Protection (Germany) suspended the registration of eight neonicotinoid pesticide seed treatment products used in oilseed rape and sweetcorn, a few weeks after honey bee keepers in the southern state of Baden Württemberg reported a wave of honey bee deaths linked to one of the pesticides, clothianidin.	Highly Toxic
Thiamethoxam	Actara	Neonicotinoid		Clothianidin is a major metabolite of Thiamethoxam. A two-year study published in 2012 showed the presence of clothianidin and thiamethoxam in bees found dead in and around hives situated near agricultural fields. Other bees at the hives exhibited tremors and uncoordinated movement and convulsions, all signs of insecticide poisoning.	Highly Toxic

Imidacloprid	Confidor, Gaucho, Kohinor, Admire, Advantage, Merit, Confidor, Hachikusan, Amigo, Seed-Plus (Chemtura Corp.), Monceren GT, Premise, Prothor, and Winner	Neonicotinoid		(Imidacloprid effects on bee population) Banned in France since 1999	highly toxic
Dicofol		Acaricide			Relatively nontoxic
Petroleum oils					Relatively nontoxic
2,4-D	ingredient in over 1,500 products	Synthetic auxin herbicide			Relatively nontoxic

Neonicotinoids

Neonicotinoids are one of the leading suspected causes of colony collapse disorder in honey bees. The specific causes are unclear, however, there has been some research to show that neonicotinoids have deleterious health effects on colony queens. Managed honeybee colonies are colonies that are "man-made." That is, they are not naturally occurring. Rather they are raised and rented out to farmers.

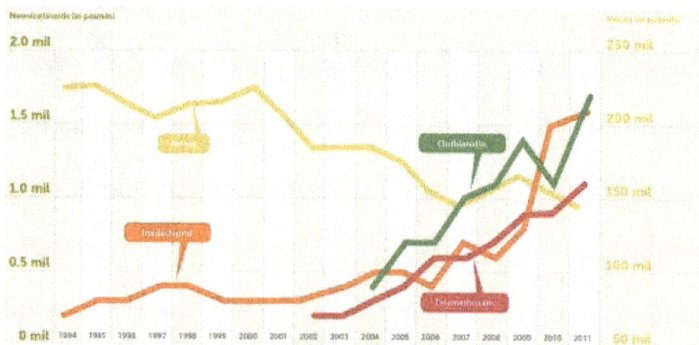

Decline of annual honey production between 2000 and 2011 by one third from 221 million to 148 million pounds with increased Neonicotinoids use in the United States.

There is some controversy surrounding the specific issue of whether or not neonicotinoids actually do negatively affect managed honeybee colonies. Perhaps one of the most popular studies showing a significant association between colony collapse disorder and neonicotinoids is 'Sub-lethal exposure to neonicotinoids impaired honeybees Winterization before proceeding to colony collapse disorder' by Chensheng Lu. Chengsheng became somewhat of a folk hero among environmental activists after his Harvard study was published, however, there has been some dissent from bee researchers in Australia, Canada, and even the USDA. That is not to say that Mr. Lu's findings are not accurate. The problem is that there simply is not a consensus yet on the real association between neonicotinoids and colony collapse disorder.

Common Insecticides Toxic to Bees and used on Soybeans

Many insecticides used against soybean aphids are highly toxic to bees.

- Orthene 75S (Acephate)
- Address 75 WSP (Acephate)
- Sevin (Carbaryl)
- Lorsban 4E, Chlorpyrifos, Eraser, Govern, Nufos, Pilot, Warhawk, Whirlwind and Yuma (Chlorpyrifos)
- Dimate (Dimethoate)
- Steward 1.25 SC (Indoxacarb)
- Lannate (Methomyl)
- Cheminova Methyl 4EC (Methyl Parathion)
- Penncap M (microencapsulated Methyl Parathion)
- Tracer (Spinosad)
- Tombstone (Cyfluthrin)
- Baythroid XL (Beta-cyfluthrin)
- Delta Gold (Deltamethrin)

Highly Toxic and Banned in the US

- Aldrin banned by US EPA in 1974
- Dieldrin banned by US EPA in 1974
- Heptachlor
- Lindane, BHC (banned in California). Lindane was also denied re-registration for agricultural use in the US by the EPA in 2006

EPA Proposal to Protect Bees from Acutely Toxic Pesticides in the US

The EPA is proposing to prohibit the application of certain pesticides and herbicides known toxic to bees during pollination periods when crops are in bloom. Growers routinely contract with honeybee keepers to bring in bees to pollinate their crops that require insect pollination. Bees are typically present during the period the crops are in bloom. Application of pesticides during this period can significantly affect the health of bees. These restrictions are expected to reduce the likelihood of high levels of pesticide exposure and mortality for bees providing pollination services. Moreover, the EPA believes these additional measures to protect bees providing pollination services will protect other pollinators as well.

The proposed restrictions would apply to all products that have liquid or dust formulations as applied, foliar use (applying pesticides directly to crop leaves) directions for use on crops, and active ingredients that have been determined via testing to have high toxicity for bees (less than 11 micrograms per bee). These restrictions would not replace already existing more restrictive, chemical-specific, and bee-protective provisions. Additionally, the proposed label restrictions would not apply to applications made in support of a government-declared public health response, such as use for wide area mosquito control. There would be no other exceptions to these proposed restrictions.

General Measures to Prevent Pesticides Bee Kills

Application of Pesticides at Evening or Night

Avoiding pesticide application directly to blooming flowers as much as possible can help limit the exposure of honeybees to toxic materials as honeybees are attracted to all types of blooming flowers. If blooming flowers must be sprayed with pesticides for any reason, they should be sprayed in the evening or night hours as bees are not in the field at that time. Usual foraging hours of honeybees are when the temperature is above 55-60 °F during the daytime, and by the evening, the bees return to the hives.

References

- Ramesh C. Gupta (28 April 2011). Toxicology of Organophosphate & Carbamate Compounds. Academic Press. pp. 352–353. ISBN 978-0-08-054310-9.

- Denis Hamilton; Stephen Crossley (14 May 2004). Pesticide Residues in Food and Drinking Water: Human Exposure and Risks. John Wiley & Sons. p. 280. ISBN 978-0-470-09160-9.

- Lewis A. Owen; Professor Kevin T Pickering; Kevin T. Pickering (1 March 2006). An Introduction to Global Environmental Issues. Routledge. p. 197. ISBN 978-1-134-76919-3.

- Astoviza, Malena J. (15 April 2014). "Evaluación de la distribución de contaminantes orgánicos persistentes (COPs) en aire en la zona de la cuenca del Plata mediante muestreadores pasivos artificiales" (in Spanish): 160. Retrieved 16 April 2014.

Significant Aspects of Environmental Toxicology

Environmental toxicology deals largely with environmental hazards that are caused by the bio-accumulation of chemical substances. This accumulation of toxins results in ecological death of many organisms or in the impairment of the organism's ability to function. The role of pathogens cannot be undermined either; hence in this chapter we find an in-depth analysis of all these issues. The aspects elucidated in this chapter are of vital importance, and provide a better understanding of environmental toxicology.

Environmental Hazard

An environmental hazard is a substance, state or event which has the potential to threaten the surrounding natural environment and / or adversely affect people's health. This term incorporates topics like pollution and natural disasters such as storms and earthquakes.

Human-made hazards while not immediately health-threatening may turn out detrimental to man's well-being eventually, because deterioration in the environment can produce secondary, unwanted negative effects on the human ecosphere. The effects of water pollution may not be immediately visible because of a sewage system that helps drain off toxic substances. If those substances turn out to be persistent (e.g. persistent organic pollutant), however, they will literally be fed back to their producers via the food chain: plankton -> edible fish -> humans. In that respect, a considerable number of environmental hazards listed below are man-made (anthropogenic) hazards.

Hazards can be categorized in four types:

1. Chemical
2. Physical (mechanical, etc.)
3. Biological
4. Psychosocial

Chemical

Chemical hazards are defined in the Globally Harmonized System and in the European Union chemical regulations. They are caused by chemical substances causing significant damage to the environment. The label is particularly applicable towards substances with aquatic toxicity. An example is zinc oxide, a common paint pigment, which is extremely toxic to aquatic life.

Toxicity or other hazards do not imply an environmental hazard, because elimination by sunlight (photolysis), water (hydrolysis) or organisms (biological elimination) neutralizes many reactive or poisonous substances. Persistence towards these elimination mechanisms combined with toxicity gives the substance the ability to do damage in the long term. Also, the lack of immediate human toxicity does not mean the substance is environmentally nonhazardous. For example, tanker truck-sized spills of substances such as milk can cause a lot of damage in the local aquatic ecosystems: the added biological oxygen demand causes rapid eutrophication, leading to anoxic conditions in the water body.

All hazards in this category are mainly anthropogenic although there exist a number of natural carcinogens and chemical elements like radon and lead may turn up in health-critical concentrations in the natural environment:

- Anthrax

- Antibiotic agents in animals destined for human consumption

- Arsenic - a contaminant of fresh water sources (water wells)

- Asbestos - carcinogenic

- DDT

- Carcinogens

- dioxins

- Endocrine disruptors

- Explosive material

- Fungicides

- Furans

- Haloalkanes

- Heavy metals

- Herbicides

- Hormones in animals destined for human consumption

- Lead in paint

- Marine debris

- mercury

- Mutagens

- Pesticides

- Polychlorinated biphenyls

- Radon and other natural sources of radioactivity

- Soil pollution
- Tobacco smoking
- Toxic waste

Physical

- Cosmic rays
- Drought
- Earthquake
- Electromagnetic fields
- E-waste
- Floods
- Fog
- Light pollution
- Lighting
- Lightning
- Noise pollution
- Quicksand
- Ultraviolet light
- vibration
- Wildfire
- X-rays

Biological

- Allergies
- Arbovirus
- Avian influenza
- Bovine spongiform encephalopathy (BSE)
- Cholera
- Ebola
- Epidemics
- Food poisoning
- Malaria

- Molds

- Onchocerciasis (river blindness)

- Pandemics

- Pathogens

- Pollen for allergic people

- Rabies

- Severe acute respiratory syndrome (SARS)

- Sick building syndrome

Ecological Death

Ecological death is the inability of an organism to function in an ecological context, leading to death. This term can be used in many fields of biology to describe any species. In the context of aquatic toxicology, a toxic chemical, or toxicant, directly affects an aquatic organism but does not immediately kill it; instead it impairs an organism's normal ecological functions which then lead to death or lack of offspring. The toxicant makes the organism unable to function ecologically in some way, even though it does not suffer obviously from the toxicant. Ecological death may be caused by sublethal toxicological effects that can be behavioral, physiological, biochemical, or histological.

Types of Sublethal Effects Causing Ecological Death

Sublethal effects consist of any effects of an organism caused by a toxicant that do not include death. These effects are generally not observed well in a shorter acute toxicity test. A longer, chronic toxicity test will allow enough time for these effects to appear in an organism and for them to lead to ecological death.

Behavioral Effects

Toxicants can affect an organism's behavior, which with aquatic organisms, may impact their ability to swim, feed or avoid predators. The impacted behavior can lead to an organism's death because it may starve or get eaten by predators. Toxicants may affect behavior by impacting the sensory systems which organisms depend on to collect information about their environment or by impacting an organism's motivation to properly respond to sensory cues. If an organism is unable to use sensory cues effectively, they may be unable to respond to early warning signs of predation risk. Toxicants can also affect later stages of predation by impacting an organism's ability to respond to predators or follow through with escape strategies.

Physiological Effects

Toxicants can affect an organism's physiology which may impact its growth, reproduction, and/ or development. If an organism does not grow correctly and is undersize or has growth defects, it will be more likely to be eaten by predators. If an organism's reproduction is impaired, it may not directly die, but it will be unable to pass on its genes to the population. The organism will no longer be representative in the population's gene pool.

Biochemical Effects

Toxicants can alter the enzymes or ions present in an organism. If this alteration does not directly cause death, but impacts the behavior or physiology of the organism, it can also lead to ecological death.

Histological Effects

Toxicants can alter an organism's tissues. If this alteration does not directly cause death, but impacts the behavior or physiology of the organism, it can also lead to ecological death.

Toxicant Examples Leading to Ecological Death

DDT

An effect caused by DDT is shell thinning in bird eggs, leading to the death of the chick. Once DDT has been accumulated by an adult bird, it is metabolized into the form DDE which is both stable and toxic. Once in the form of DDE, the chemical impacts the metabolism of calcium in adult female birds' shell glands, ultimately causing a decrease in eggshell thickness. At high concentrations of DDT, the eggshells will no longer be able to support the incubating parents' weight and will lead to the death of the unborn chick. This is an example of physiological and biochemical sublethal effects leading to ecological death of the chick.

Diazinon

An effect caused by diazinon is a decrease in response to predator cues in Chinook salmon (*Oncorhynchus tshawytscha*). Diazinon, an organopesticide, was exposed to juvenile Chinook salmon for two hours at 1 and 10 µg/L, and these concentrations were enough to eliminate the behavioral responses of the fish to predator chemical cues. If the fish cannot recognize that a predator is nearby, it is likely to be eaten. This is an example of a behavioral sublethal effect leading to ecological death.

Pentachlorophenol

An effect caused by pentachlorophenol is a decrease in response to predator attacks in guppies (*Poecilia reticula*). Pentachlorophenol was exposed to juvenile guppies at 500 and 700 µg/L, and both concentrations decreased the guppies' reactions to predators. The predators did not have to strike as frequently, did not have to pursue as frequently, or have to pursue the guppies as long as guppies that had not been exposed to these levels of pentachlorophenol. The guppies that were exposed to this chemical were more likely to be eaten due to their slower responses. This is another example of a behavioral sublethal effect that leads to ecological death.

Copper

An effect caused by copper is impacting the salmon olfactory system. The olfactory system is used to gather important information about one's environment. In the case of salmon, olfactory cues can gather information about habitat quality, predators, mates and more. Salmon can detect distinct copper gradients using their olfactory system, and use this information to avoid contaminated waters. However, when salmon are affected by copper, the olfactory system can be impacted in a matter of minutes. If the fish is no longer able to gather environmental information, it may be at risk for predation or unable to find resources for survival. This is an example of a physiological sublethal effect leading to ecological death.

Persistent, Bioaccumulative and Toxic Substances

Persistent, bioaccumulative and toxic (PBTs) substances are a class of compounds that have high resistance to degradation from abiotic and biotic factors, high mobility in the environment and high toxicity. Because of these factors PBTs have been observed to have a high order of bioaccumulation and biomagnification, very long retention times in various media, and widespread distribution across the globe. Majority of PBTs in the environment are either created through industry or are unintentional byproducts.

History

Persistent organic pollutants (POPs) were the focal point of the Stockholm Convention 2001 due to their persistence, ability to biomagnify and the threat posed to both human health and the environment. The goal of the Stockholm Convention was to determine the classification of POPs, create measures to eliminate production/use of POPs, and establish proper disposal of the compounds in an environmentally friendly manner. Currently the majority of the global community is actively involved with this program but a few still resist, most notably the U.S.

Similar to POPs classification, the PBT classification of chemicals was developed in 1997 by the Great Lakes Binational Toxic Strategy (GLBNS). Signed by both the U.S and Canada, the GLBNS classified PBTs in one of two categories, level I and level II. Level I PBTs are top priority which currently, as of 2005, contained 12 compounds or classes of compounds.

Level I PBTs (GLBNS)

- Mercury
- Polychlorinated biphenyls PCBs)
- Dioxins/furans
- Benzo(a)pyrene (BaP)
- Hexachlorobenzene (HCB)
- Alkyl-lead

- Pesticides
    - Mirex
    - Dieldrin/aldrin
    - Chlordane
    - Toxaphene
- Octachlorostyrene

The GLBNS is administered by the U.S Environmental Protection Agency (USEPA) and Environment Canada. Following the GLBNS, the Multimedia Strategy for Priority Persistent, Bioaccumulative and Toxic Pollutants (PBT Strategy) was drafted by the USEPA. The PBT Strategy led to the implementation of PBT criteria in several regulational policies. Two main policies that were changed by the PBT strategy were the Toxics Release Inventory (TRI) which required more rigid chemical reporting and the New Chemical Program (NCP) under the Toxics Substances Control Act (TSCA) which required screening for PBTs and PBT properties.

Compounds

General

PBTs are a unique classification of chemicals that have and will continue to impact human health and the environment worldwide. The three main attributes of a PBT (persistence, bioaccumulative and toxic) each have a huge role in the risk posed by these compounds.

Persistence

PBTs have a high environmental mobility relative to other contaminants mainly due to their resistance to degradation (persistence). This allows PBTs to travel far and wide in both the atmosphere and in aqueous environments. The low degradation rates of PBTs allow these chemicals to be exposed to both biotic and abiotic factors while maintaining a relatively stable concentration. Another factor that makes PBTs especially dangerous are the degradation products which are often relatively as toxic as the parent compound. These factors have resulted in global contamination most notable in remote areas such as the arctic and high elevation areas which are far from any source of PBTs.

Bioaccumulation and Biomagnification

The bioaccumulative ability of PBTs follows suit with the persistence attribute by the high resistance to degradation by biotic factors, especially with in organisms. Bioaccumulation is the result of a toxic substance being taken up at a higher rate than being removed from an organism. For PBTs this is caused mainly by a resistance to degradation, biotic and abiotic. PBTs usually are highly insoluble in water which allows them to enter organisms at faster rates through fats and other nonpolar regions on an organism. Bioaccumulation of a toxicant can lead to biomagnification through a trophic web which has resulted in massive concern in areas with especially low trophic diversity. Biomagnification results in higher trophic organisms accumulating more PBTs than those of lower trophic levels through consumption of the PBT contaminated lower trophic organisms.

Toxicity

The toxicity of this class of compounds is high with very low concentrations of a PBT required to enact an effect on an organism compared to most other contaminants. This high toxicity along with the persistence allows for the PBT to have detrimental effects in remote areas around the globe where there is not a local source of PBTs. The bioaccumulation and magnification along with the high toxicity and persistence has the ability to destroy and/or irreparably damage trophic systems, especially the higher trophic levels, globally. It is this reason that PBTs have become an area of focus in global politics.

Specific Toxicants

PCBs

Historically, PCBs were used extensively for industrial purposes such as coolants, insulating fluids, and as a plasticizer. These contaminants enter the environment through both use and disposal. Due to extensive concern from the public, legal, and scientific sectors indicating that PCBs are likely carcinogens and potential to adversely impact the environment, these compounds were banned in 1979 in the United States. The ban included the use of PCBs in uncontained sources, such as adhesives, fire retardant fabric treatments, and plasticizers in paints and cements. Containers that are completely enclosed such as transformers and capacitors are exempt from the ban.

The inclusion of PCBs as a PBT may be contributed to their low water solubility, high stability, and semi-volatility facilitating their long range transport and accumulation in organisms. The persistence of these compounds is due to the high resistance to oxidation, reduction, addition, elimination and electrophilic substitution. The toxicological interactions of PCBs are affected by the number and position of the chlorine atoms, without ortho substitution are referred as coplanar and all others as non-coplanar. Non-coplanar PCBs may cause neurotoxicity by interfering with intracellular signal transduction dependent on calcium. Ortho-PCBs may alter hormone regulation through disruption of the thyroid hormone transport by binding to transthyretin. Coplanar PCBs are similar to dioxins and furans, both bind to the aryl hydrocarbon receptor (AhR) in organisms and may exert dioxin-like effects, in addition to the effects shared with non-coplanar PCBs. The AhR is a transcription factor, therefore, abnormal activation may disrupt cellular function by altering gene transcription.

Effects of PBTs may include increase in disease, lesions in benthic feeders, spawning loss, change in age-structured populations of fish, and tissue contamination in fish and shellfish. Humans and other organisms, which consume shellfish and/or fish contaminated with persistent bioaccumulative pollutants, have the potential to bioaccumulate these chemicals. This may put these organisms at risk of mutagenic, teratogenic, and/or carcinogenic effects. Correlations have been found between elevated exposure to PCB mixtures and alterations in liver enzymes, hepatomegaly, and dermatological effects such as rashes have been reported.

DDT

One PBT of concern includes DDT (dichlorodiphenyltrichloroethane), an organochlorine that was widely used as an insecticide during World War II to protect soldiers from malaria carried by mosquitoes. Due to the low cost and low toxicity to mammals, the widespread use of DDT for agricultural and commercial motives started around 1940. However, the overuse of DDT lead to insect tolerance to the chemical. It was also discovered that DDT had a high toxicity to fish. DDT was banned in the US by 1973 because of building evidence that DDT's stable structure, high fat solubility, and low rate of metabolism, caused it to bioaccumulate in animals. While DDT is banned in the US, other countries such as China and Turkey still produce and use it quite regularly through Dicofol, an insecticide that has DDT as an impurity. This continued use in other parts of the world is still a global problem due to the mobility and persistence of DDT.

The initial contact from DDT is on vegetation and soil. From here, the DDT can travel many routes, for instance, when plants and vegetation are exposed to the chemical to protect from insects, the plants may absorb it. Then these plants may either be consumed by humans or other animals. These consumers ingest the chemical and begin metabolizing the toxicant, accumulating more through ingestion, and posing health risks to the organism, their offspring, and any predators. Alternatively the ingestion of the contaminated plant by insects may lead to tolerance by the organism. Another route is the chemical travelling through the soil and ending up in ground water and in human water supply. Or in the case that the soil is near a moving water system, the chemical could end up in large freshwater systems or the ocean where fish are at high risk from the toxicological effects of DDT. Lastly, the most common transport route is the evaporation of DDT into the atmosphere followed by condensation and eventually precipitation where it is released into environments anywhere on earth. Due to the long-range transport of DDT, the presence of this harmful toxicant will continue as long as it is still used anywhere and until the current contamination eventually degrades. Even after its complete discontinued use, it will still remain in the environment for many more years after because of DDT's persistent attributes.

Previous studies have shown that DDT and other similar chemicals directly elicited a response and effect from excitable membranes. DDT causes membranes such as sense organs and nerves endings to activate repetitively by slowing down the ability for the sodium channel to close and stop releasing sodium ions. The sodium ions are what polarize the opposing synapse after it has depolarized from firing. This inhibition of closing the sodium ion channel can lead to a variety of problems including a dysfunctional nervous system, decreased motor abilities/function/control, reproductive impairment (egg-shell thinning in birds), and development deficiencies. Presently, DDT has been labeled as a possible human carcinogen based on animal liver tumor studies. DDT toxicity on humans have been associated with dizziness, tremors, irritability, and convulsions. Chronic toxicity has led to long term neurological and cognitive issues.

Mercury

Inorganic

Inorganic mercury (elemental mercury) is less bioavailable and less toxic than that of organic mercury but is still toxic nonetheless. It is released into the environment through both natural sources as well as human activity, and it has the capability to travel long distances through the atmosphere.

Around 2,700 to 6,000 tons of elemental mercury are released via natural activity such as volcanoes and erosion. Another 2,000 – 3,000 tons are released by human industrial activities such as coal combustion, metal smelting and cement production. Due to its high volatility and atmospheric residence time of around 1 year, mercury has the ability to travel across continents before being deposited. Inorganic mercury has a wide spectrum of toxicological effects that include damage to the respiratory, nervous, immune and excretory systems in humans. Inorganic mercury also possesses the ability to bioaccumulate individuals and biomagnify through trophic systems.

Organic

Organic mercury is significantly more detrimental to the environment than its inorganic form due to its widespread distribution as well as its higher mobility, general toxicity and rates of bioaccumulation than that of the inorganic form. Environmental organic mercury is mainly created by the transformation of elemental (inorganic) mercury via anaerobic bacteria into methylated mercury (organic). The global distribution of organic mercury is the result of general mobility of the compound, activation via bacteria and transportation from animal consumption. Organic mercury shares a lot of the same effects as the inorganic form but it has a higher toxicity due to its higher mobility in the body, especially its ability to readily move across the blood brain barrier.

Ecological Impact of Hg

The high toxicity of both forms of mercury (especially organic mercury) poses a threat to almost all organisms that comes in contact with it. This is one of the reasons that there is such high attention to mercury in the environment but even more so than its toxicity is both its persistence and atmospheric retention times. The ability of mercury to readily volatilize allows it to enter the atmosphere and travel far and wide. Unlike most other PBTs that have atmospheric half-lives between 30 min and 7 days mercury has an atmospheric residence time of at least 1 year. This atmospheric retention time along with mercury's resistance to degradation factors such as electromagnetic radiation and oxidation, which are two of the main factors leading to degradation of many PBTs in the atmosphere, allows mercury from any source to be transported extensively. This characteristic of mercury transportation globally along with its high toxicity is the reasoning behind its incorporation into the BNS list of PBTs.

Notable PBT environmental impacts

Japan

The realization of the adverse effects from environmental pollution were made public from several disasters that occurred globally. In 1965, it was recognized that extensive mercury pollution by the Chisso chemical factory in Minamata, Japan due to improper handling of industrial wastes resulted in significant effects to the humans and organisms exposed. Mercury was released into the environment as methyl mercury (bioavailable state) into industrial wastewater and was then bioaccumulated by shellfish and fish in Minamata Bay and the Shiranui Sea. When the contaminated seafood was consumed by the local populace it caused a neurological syndrome, coined Minamata disease. Symptoms include general muscle weakness, hearing damage, reduced field of vision, and ataxia. The Minamata disaster contributed to the global realization of the potential dangers from environmental pollution and to the characterization of PBTs.

Puget Sound

Despite the ban on DDT 30 years earlier and years of various efforts to clean up Puget Sound from DDT and PCB's, there is still a significant presence of both compounds which pose a constant threat to human health and the environment. Harbor seals (*Phoca vitulina*), a common marine species in the Puget Sound area have been the focus of a few studies to monitor and examine the effects of DDT accumulation and magnification in aquatic wildlife. One study tagged and reexamined seal pups every 4 to 5 years to be tested for DDT concentrations. The trends showed the pups to be highly contaminated; this means their prey are also highly contaminated. Due to DDT's high lipid solubility, it also has the ability to accumulate in the local populace who consume seafood from the area. This also translates to women who are pregnant or breastfeeding, since DDT will be transferred from the mother to child. Both animal and human health risk to DDT will continue to be an issue in Puget Sound especially because of the cultural significance of fish in this region.

Pathogen

In biology, a pathogen (pathos "suffering, passion" and -genēs "producer of") in the oldest and broadest sense is anything that can produce disease; the term came into use in the 1880s. Typically the term is used to describe an infectious agent such as a virus, bacterium, prion, a fungus, or even another micro-organism.

There are several substrates including *pathways* where the pathogens can invade a host. The principal pathways have different episodic time frames, but soil contamination has the longest or most persistent potential for harboring a pathogen. Diseases caused by organisms in humans are known as pathogenic diseases.

Pathogenicity

Pathogenicity is the potential disease-causing capacity of pathogens. Pathogenicity is related to virulence in meaning, but some authorities have come to distinguish it as a *qualitative* term, whereas the latter is *quantitative*. By this standard, an organism may be said to be pathogenic or non-pathogenic in a particular context, but not "more pathogenic" than another. Such comparisons are described instead in terms of relative virulence. Pathogenicity is also distinct from the transmissibility of the virus, which quantifies the risk of infection.

A pathogen may be described in terms of its ability to produce toxins, enter tissue, colonize, hijack nutrients, and its ability to immunosuppress the host.

Context-dependent Pathogenicity

It is common to speak of an entire species of bacteria as pathogenic when it is identified as the cause of a disease *(cf. Koch's postulates)*. However, the modern view is that pathogenicity depends on the microbial ecosystem as a whole. A bacterium may participate in opportunistic infections in immunocompromised hosts, acquire virulence factors by plasmid infection, become transferred to a different site within the host, or respond to changes in the overall numbers of other bacteria present. For example, infection of mesenteric lymph glands of mice with *Yersinia* can clear the way for continuing infection of these sites by *Lactobacillus*, possibly by a mechanism of "immunological scarring".

Related concepts

Virulence

Virulence (the tendency of a pathogen to cause damage to a host's fitness) evolves when that pathogen can spread from a diseased host, despite that host being very debilitated. Horizontal transmission occurs between hosts of the same species, in contrast to vertical transmission, which tends to evolve symbiosis (after a period of high morbidity and mortality in the population) by linking the pathogen's evolutionary success to the evolutionary success of the host organism.

Evolutionary medicine has found that under horizontal transmission, the host population might never develop tolerance to the pathogen.

Transmission

Transmission of pathogens occurs through many different routes, including airborne, direct or indirect contact, sexual contact, through blood, breast milk, or other body fluids, and through the fecal-oral route.

Types of pathogens

Bacterial

Although the vast majority of bacteria are harmless or beneficial, a relatively small list of pathogenic bacteria can cause infectious diseases. One of the bacterial diseases with the highest disease burden is tuberculosis, caused by the bacterium *Mycobacterium tuberculosis*, which kills about 2 million people a year, mostly in sub-Saharan Africa. Pathogenic bacteria contribute to other globally important diseases, such as pneumonia, which can be caused by bacteria such as *Streptococcus* and *Pseudomonas*, and foodborne illnesses, which can be caused by bacteria such as *Shigella*, *Campylobacter*, and *Salmonella*. Pathogenic bacteria also cause infections such as tetanus, typhoid fever, diphtheria, syphilis, and leprosy.

Bacteria can often be killed by antibiotics because the cell wall on the outside is destroyed, expelling the DNA out of the body of the pathogen, therefore making the pathogen incapable of producing proteins and dies. Bacteria typically range between 1 and 5 micrometers in length. A class of bacteria without cell walls is mycoplasma (a cause of lung infections). A class of bacteria which must live within other cells (obligate intracellular parasitic) is chlamydia (genus), the world leader in causing sexually transmitted infection (STD).

Viral

Some of the diseases that are caused by viral pathogens include smallpox, influenza, mumps, measles, chickenpox, ebola, and rubella.

Pathogenic viruses are diseases mainly those of the families of: Adenoviridae, Picornaviridae, Herpesviridae, Hepadnaviridae, Flaviviridae, Retroviridae, Orthomyxoviridae, Paramyxoviridae, Papovaviridae, Polyomavirus, Rhabdoviridae, Togaviridae. Viruses typically range between 20-300 nanometers in length.

Fungal

Fungi comprise a eukaryotic kingdom of microbes that are usually saprophytes (consume dead organisms) but can cause diseases in humans, animals and plants. Fungi are the most common cause of diseases in crops and other plants. The typical fungal spore size is 1-40 micrometers in length.

Prionic

According to the prion theory, prions are infectious pathogens that do not contain nucleic acids. These abnormally folded proteins are found characteristically in some diseases such as scrapie, bovine spongiform encephalopathy (mad cow disease) and Creutzfeldt–Jakob disease.

Other Parasites

Some eukaryotic organisms, such as protists and helminths, cause disease.

Treatment and Health Care

Bacteria are usually treated with antibiotics while viruses are treated with antiviral compounds. Eukaryotic pathogens are typically not susceptible to antibiotics and thus need specific drugs. Infection with many pathogens can be prevented by immunization. A small amount of pathogens are used in vaccines to make immunity stay alert and strengthen defense on the insides to prepare for a larger quantity of the virus ever getting inside. Hygiene is critical for the prevention of infection by pathogens.

References

- Alberts B; Johnson A; Lewis J; et al. (2002). "Introduction to Pathogens". Molecular Biology of the Cell (4th ed.). Garland Science. p. 1. Retrieved 26 April 2016.

- Carl Nathan (2015-10-09). "From transient infection to chronic disease". Science. 350 (6257): 161. doi:10.1126/science.aad4141. PMID 26450196.

- Casadevall, Arturo; Pirofski, Liise-anne (11 December 2014). "Ditch the term pathogen". Comment. Nature (paper). 516 (7530): 165–6. doi:10.1038/516165a.

- USEPA. The Great Lakes Water Quality Agreement U.S Eighth Response to International Joint Commission. Retrieved June 6, 2012

Evaluation Methods of Toxicology

To determine toxicity levels of the environment, there exist several measuring scales that enable a statistical understanding. This chapter of the book talks about scales and indexes like- dietary reference intake, air quality index, early life stage test etc. The chapter also studies the modes of toxic action and the quantification of toxic substances in living organisms. This chapter discusses the methods of environmental toxicology in a critical manner providing key analysis to the subject matter.

Air Quality Index

An air quality index (AQI) is a number used by government agencies to communicate to the public how polluted the air currently is or how polluted it is forecast to become. As the AQI increases, an increasingly large percentage of the population is likely to experience increasingly severe adverse health effects. Different countries have their own air quality indices, corresponding to different national air quality standards. Some of these are the Air Quality Health Index (Canada), the Air Pollution Index (Malaysia), and the Pollutant Standards Index (Singapore).

Wildfires give rise to an elevated AQI in parts of Greece

Definition and Usage

An air quality measurement station in Edinburgh, Scotland

Computation of the AQI requires an air pollutant concentration over a specified averaging period, obtained from an air monitor or model. Taken together, concentration and time represent the dose of the air pollutant. Health effects corresponding to a given dose are established by epidemiological research. Air pollutants vary in potency, and the function used to convert from air pollutant concentration to AQI varies by pollutant. Air quality index values are typically grouped into ranges. Each range is assigned a descriptor, a color code, and a standardized public health advisory.

The AQI can increase due to an increase of air emissions (for example, during rush hour traffic or when there is an upwind forest fire) or from a lack of dilution of air pollutants. Stagnant air, often caused by an anticyclone, temperature inversion, or low wind speeds lets air pollution remain in a local area, leading to high concentrations of pollutants, chemical reactions between air contaminants and hazy conditions.

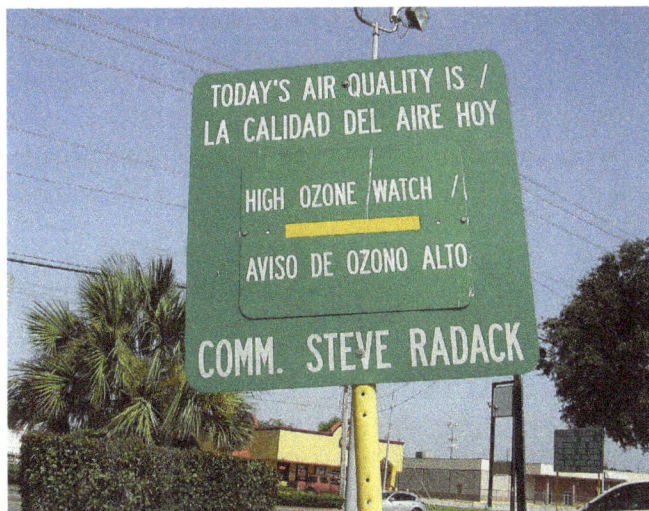

Signboard in Gulfton, Houston indicating an ozone watch

On a day when the AQI is predicted to be elevated due to fine particle pollution, an agency or public health organization might:

- advise sensitive groups, such as the elderly, children, and those with respiratory or cardio-vascular problems to avoid outdoor exertion.

- declare an "action day" to encourage voluntary measures to reduce air emissions, such as using public transportation.

- recommend the use of masks to keep fine particles from entering the lungs

Woman wearing an air pollution mask in Beijing, China

During a period of very poor air quality, such as an air pollution episode, when the AQI indicates that acute exposure may cause significant harm to the public health, agencies may invoke emergency plans that allow them to order major emitters (such as coal burning industries) to curtail emissions until the hazardous conditions abate.

Most air contaminants do not have an associated AQI. Many countries monitor ground-level ozone, particulates, sulfur dioxide, carbon monoxide and nitrogen dioxide, and calculate air quality indices for these pollutants.

The definition of the AQI in a particular nation reflects the discourse surrounding the development of national air quality standards in that nation. A website allowing government agencies anywhere in the world to submit their real-time air monitoring data for display using a common definition of the air quality index has recently become available.

Indices by Location

Canada

Air quality in Canada has been reported for many years with provincial Air Quality Indices (AQIs). Significantly, AQI values reflect air quality management objectives, which are based on the lowest achievable emissions rate, and not exclusively concern for human health. The Air Quality Health Index or (AQHI) is a scale designed to help understand the impact of air quality on health. It is a health protection tool used to make decisions to reduce short-term exposure to air pollution by adjusting activity levels during increased levels of air pollution. The Air Quality Health Index also provides advice on how to improve air quality by proposing behavioural change to reduce the environmental footprint. This index pays particular attention to people who are sensitive to air pollution. It provides them with advice on how to protect their health during air quality levels associated with low, moderate, high and very high health risks.

The Air Quality Health Index provides a number from 1 to 10+ to indicate the level of health risk associated with local air quality. On occasion, when the amount of air pollution is abnormally high, the number may exceed 10. The AQHI provides a local air quality current value as well as a local air quality maximums forecast for today, tonight, and tomorrow, and provides associated health advice.

| 1 | 2 | 3 | 4 | 5 | 6 | 7 | 8 | 9 | 10 | + |

| Risk: | Low (1–3) | Moderate (4–6) | High (7–10) | Very high (above 10) |

Health Risk	Air Quality Health Index	Health Messages	
		At Risk population	*General Population
Low	1–3	**Enjoy** your usual outdoor activities.	**Ideal** air quality for outdoor activities
Moderate	4–6	**Consider reducing** or rescheduling strenuous activities outdoors if you are experiencing symptoms.	**No need to modify** your usual outdoor activities unless you experience symptoms such as coughing and throat irritation.
High	7–10	**Reduce** or reschedule strenuous activities outdoors. Children and the elderly should also take it easy.	**Consider reducing** or rescheduling strenuous activities outdoors if you experience symptoms such as coughing and throat irritation.
Very high	Above 10	**Avoid** strenuous activities outdoors. Children and the elderly should also avoid outdoor physical exertion.	**Reduce** or reschedule strenuous activities outdoors, especially if you experience symptoms such as coughing and throat irritation.

Hong Kong

On the 30th December 2013 Hong Kong replaced the Air Pollution Index with a new index called the *Air Quality Health Index.* This index is on a scale of 1 to 10+ and considers four air pollutants: ozone; nitrogen dioxide; sulphur dioxide and particulate matter (including PM10 and PM2.5). For any given hour the AQHI is calculated from the sum of the percentage excess risk of daily hospital admissions attributable to the 3-hour moving average concentrations of these four pollutants. The AQHIs are grouped into five AQHI health risk categories with health advice provided:

Health risk category	AQHI
Low	1
	2
	3
Medium	4
	5
	6
High	7
Very High	8
	9
	10
	10+

Each of the health risk categories has advice with it. At the *low* and *moderate* levels the public are advised that they can continue normal activities. For the *high* category, children, the elderly and people with heart or respiratory illnesses are advising to reduce outdoor physical exertion. Above this (*very high* or *serious*) the general public are also advised to reduce or avoid outdoor physical exertion.

Mainland China

China's Ministry of Environmental Protection (MEP) is responsible for measuring the level of air pollution in China. As of 1 January 2013, MEP monitors daily pollution level in 163 of its major cities. The API level is based on the level of 6 atmospheric pollutants, namely sulfur dioxide (SO_2), nitrogen dioxide (NO_2), suspended particulates smaller than 10 μm in aerodynamic diameter (PM_{10}), suspended particulates smaller than 2.5 μm in aerodynamic diameter ($PM_{2.5}$CALcarbon monoxide (CO), and ozone (O_3) measured at the monitoring stations throughout each city.

AQI Mechanics

An individual score (IAQI) is assigned to the level of each pollutant and the final AQI is the highest of those 6 scores. The pollutants can be measured quite differently. $PM_{2.5}$、PM_{10} concentration are measured as average per 24h. SO_2, NO_2, O_3, CO are measured as average per hour. The final API value is calculated per hour according to a formula published by the MEP.

The scale for each pollutant is non-linear, as is the final AQI score. Thus an AQI of 100 does not mean twice the pollution of AQI at 50, nor does it mean twice as harmful. While an AQI of 50 from day 1 to 182 and AQI of 100 from day 183 to 365 does provide an annual average of 75, it does *not* mean the pollution is acceptable even if the benchmark of 100 is deemed safe. This is because the benchmark is a 24-hour target. The annual average must match against the annual target. It is entirely possible to have safe air every day of the year but still fail the annual pollution benchmark.

AQI and Health Implications (HJ 663-2012)

AQI	Air Pollution Level	Health Implications
0–50	Excellent	No health implications.
51–100	Good	Few hypersensitive individuals should reduce outdoor exercise.
101–150	Lightly Polluted	Slight irritations may occur, individuals with breathing or heart problems should reduce outdoor exercise.
151–200	Moderately Polluted	Slight irritations may occur, individuals with breathing or heart problems should reduce outdoor exercise.
201–300	Heavily Polluted	Healthy people will be noticeably affected. People with breathing or heart problems will experience reduced endurance in activities. These individuals and elders should remain indoors and restrict activities.
300+	Severely Polluted	Healthy people will experience reduced endurance in activities. There may be strong irritations and symptoms and may trigger other illnesses. Elders and the sick should remain indoors and avoid exercise. Healthy individuals should avoid outdoor activities.

India

The Minister for Environment, Forests & Climate Change Shri Prakash Javadekar launched The National Air Quality Index (AQI) in New Delhi on 17 September 2014 under the Swachh Bharat Abhiyan. It is outlined as 'One Number- One Colour-One Description' for the common man to judge the air quality within his vicinity. The index constitutes part of the Government's mission to introduce the culture of cleanliness. Institutional and infrastructural measures are being undertaken in order to ensure that the mandate of cleanliness is fulfilled across the country and the Ministry of Environment, Forests & Climate Change proposed to discuss the issues concerned regarding quality of air with the Ministry of Human Resource Development in order to include this issue as part of the sensitisation programme in the course curriculum.

While the earlier measuring index was limited to three indicators, the current measurement index had been made quite comprehensive by the addition of five additional parameters. Under the current measurement of air quality there are 8 parameters . The initiatives undertaken by the Ministry recently aimed at balancing environment and conservation and development as air pollution has been a matter of environmental and health concerns, particularly in urban areas.

The Central Pollution Control Board along with State Pollution Control Boards has been operating National Air Monitoring Program (NAMP) covering 240 cities of the country having more than 342 monitoring stations. In addition, continuous monitoring systems that provide data on near real-time basis are also installed in a few cities. They provide information on air quality in public domain in simple linguistic terms that is easily understood by a common person. Air Quality Index (AQI) is one such tool for effective dissemination of air quality information to people. As such an Expert Group comprising medical professionals, air quality experts, academia, advocacy groups, and SPCBs was constituted and a technical study was awarded to IIT Kanpur. IIT Kanpur and the Expert Group recommended an AQI scheme in 2014.

There are six AQI categories, namely Good, Satisfactory, Moderately polluted, Poor, Very Poor, and Severe. The proposed AQI will consider eight pollutants (PM_{10}, $PM_{2.5}$, NO_2, SO_2, CO, O_3, NH_3, and Pb) for which short-term (up to 24-hourly averaging period) National Ambient Air Quality Standards are prescribed. Based on the measured ambient concentrations, corresponding standards and likely health impact, a sub-index is calculated for each of these pollutants. The worst sub-index reflects overall AQI. Associated likely health impacts for different AQI categories and pollutants have been also been suggested, with primary inputs from the medical expert members of the group. The AQI values and corresponding ambient concentrations (health breakpoints) as well as associated likely health impacts for the identified eight pollutants are as follows:

AQI Category, Pollutants and Health Breakpoints

AQI Category (Range)	PM_{10} (24hr)	$PM_{2.5}$ (24hr)	NO_2 (24hr)	O_3 (8hr)	CO (8hr)	SO_2 (24hr)	NH_3 (24hr)	Pb (24hr)
Good (0-50)	0-50	0-30	0-40	0-50	0-1.0	0-40	0-200	0-0.5
Satisfactory (51-100)	51-100	31-60	41-80	51-100	1.1-2.0	41-80	201-400	0.5-1.0
Moderately polluted (101-200)	101-250	61-90	81-180	101-168	2.1-10	81-380	401-800	1.1-2.0
Poor (201-300)	251-350	91-120	181-280	169-208	10-17	381-800	801-1200	2.1-3.0
Very poor (301-400)	351-430	121-250	281-400	209-748	17-34	801-1600	1200-1800	3.1-3.5
Severe (401-500)	430+	250+	400+	748+	34+	1600+	1800+	3.5+

AQI	Associated Health Impacts
Good (0-50)	Minimal impact
Satisfactory (51-100)	May cause minor breathing discomfort to sensitive people.
Moderately polluted (101–200)	May cause breathing discomfort to people with lung disease such as asthma, and discomfort to people with heart disease, children and older adults.
Poor (201-300)	May cause breathing discomfort to people on prolonged exposure, and discomfort to people with heart disease.
Very poor (301-400)	May cause respiratory illness to the people on prolonged exposure. Effect may be more pronounced in people with lung and heart diseases.
Severe (401-500)	May cause respiratory impact even on healthy people, and serious health impacts on people with lung/heart disease. The health impacts may be experienced even during light physical activity.

Mexico

The air quality in Mexico City is reported in IMECAs. The IMECA is calculated using the measurements of average times of the chemicals ozone (O_3), sulphur dioxide (SO_2), nitrogen dioxide (NO_2), carbon monoxide (CO), particles smaller than 2.5 micrometers ($PM_{2.5}$), and particles smaller than 10 micrometers (PM_{10}).

Singapore

Singapore uses the Pollutant Standards Index to report on its air quality, with details of the calculation similar but not identical to that used in Malaysia and Hong Kong The PSI chart below is grouped by index values and descriptors, according to the National Environment Agency.

PSI	Descriptor	General Health Effects
0–50		None
51–100	Moderate	Few or none for the general population
101–200	Unhealthy	Mild aggravation of symptoms among susceptible persons i.e. those with underlying conditions such as chronic heart or lung ailments; transient symptoms of irritation e.g. eye irritation, sneezing or coughing in some of the healthy population.
201–300	Very Unhealthy	Moderate aggravation of symptoms and decreased tolerance in persons with heart or lung disease; more widespread symptoms of transient irritation in the healthy population.
301–400	Hazardous	Early onset of certain diseases in addition to significant aggravation of symptoms in susceptible persons; and decreased exercise tolerance in healthy persons.
Above 400	Hazardous	PSI levels above 400 may be life-threatening to ill and elderly persons. Healthy people may experience adverse symptoms that affect normal activity.

South Korea

The Ministry of Environment of South Korea uses the Comprehensive Air-quality Index (CAI) to describe the ambient air quality based on the health risks of air pollution. The index aims to help the public easily understand the air quality and protect people's health. The CAI is on a scale from 0 to 500, which is divided into six categories. The higher the CAI value, the greater the level of air pollution. Of values of the five air pollutants, the highest is the CAI value. The index also has associated health effects and a colour representation of the categories as shown below.

CAI	Description	Health Implications
0–50	Good	A level that will not impact patients suffering from diseases related to air pollution.
51–100	Moderate	A level that may have a meager impact on patients in case of chronic exposure.
101–150	Unhealthy for sensitive groups	A level that may have harmful impacts on patients and members of sensitive groups.
151–250	Unhealthy	A level that may have harmful impacts on patients and members of sensitive groups (children, aged or weak people), and also cause the general public unpleasant feelings.
251–500	Very unhealthy	A level that may have a serious impact on patients and members of sensitive groups in case of acute exposure.

The N Seoul Tower on Namsan Mountain in central Seoul, South Korea, is illuminated in blue, from sunset to 23:00 and 22:00 in winter, on days where the air quality in Seoul is 45 or less. During the spring of 2012, the Tower was lit up for 52 days, which is four days more than in 2011.

United Kingdom

The most commonly used air quality index in the UK is the *Daily Air Quality Index* recommended by the Committee on Medical Effects of Air Pollutants (COMEAP). This index has ten points, which are further grouped into 4 bands: low, moderate, high and very high. Each of the bands comes with advice for at-risk groups and the general population.

Air pollution banding	Value	Health messages for At-risk individuals	Health messages for General population
Low	1–3	Enjoy your usual outdoor activities.	Enjoy your usual outdoor activities.
Moderate	4–6	Adults and children with lung problems, and adults with heart problems, who experience symptoms, should consider reducing strenuous physical activity, particularly outdoors.	Enjoy your usual outdoor activities.
High	7–9	Adults and children with lung problems, and adults with heart problems, should reduce strenuous physical exertion, particularly outdoors, and particularly if they experience symptoms. People with asthma may find they need to use their reliever inhaler more often. Older people should also reduce physical exertion.	Anyone experiencing discomfort such as sore eyes, cough or sore throat should consider reducing activity, particularly outdoors.
Very High	10	Adults and children with lung problems, adults with heart problems, and older people, should avoid strenuous physical activity. People with asthma may find they need to use their reliever inhaler more often.	Reduce physical exertion, particularly outdoors, especially if you experience symptoms such as cough or sore throat.

The index is based on the concentrations of 5 pollutants. The index is calculated from the concentrations of the following pollutants: Ozone, Nitrogen Dioxide, Sulphur Dioxide, PM2.5 (particles with an aerodynamic diameter less than 2.5 μm) and PM10. The breakpoints between index values are defined for each pollutant separately and the overall index is defined as the maximum value of the index. Different averaging periods are used for different pollutants.

Index	Ozone, Running 8 hourly mean ($\mu g/m^3$)	Nitrogen Dioxide, Hourly mean ($\mu g/m^3$)	Sulphur Dioxide, 15 minute mean ($\mu g/m^3$)	PM2.5 Particles, 24 hour mean ($\mu g/m^3$)	PM10 Particles, 24 hour mean ($\mu g/m^3$)
1	0-33	0-67	0-88	0-11	0-16
2	34-66	68-134	89-177	12-23	17-33
3	67-100	135-200	178-266	24-35	34-50
4	101-120	201-267	267-354	36-41	51-58
5	121-140	268-334	355-443	42-47	59-66
6	141-160	335-400	444-532	48-53	67-75

7	161-187	401-467	533-710	54-58	76-83
8	188-213	468-534	711-887	59-64	84-91
9	214-240	535-600	888-1064	65-70	92-100
10	≥ 241	≥ 601	≥ 1065	≥ 71	≥ 101

Europe

To present the air quality situation in European cities in a comparable and easily understandable way, all detailed measurements are transformed into a single relative figure: the Common Air Quality Index (or CAQI) Three different indices have been developed by Citeair to enable the comparison of three different time scale:.

- An hourly index, which describes the air quality today, based on hourly values and updated every hours,

- A daily index, which stands for the general air quality situation of yesterday, based on daily values and updated once a day,

- An annual index, which represents the city's general air quality conditions throughout the year and compare to European air quality norms. This index is based on the pollutants year average compare to annual limit values, and updated once a year.

However, the proposed indices and the supporting common web site www.airqualitynow.eu are designed to give a dynamic picture of the air quality situation in each city but not for compliance checking.

The Hourly and Daily Common Indices

These indices have 5 levels using a scale from 0 (very low) to > 100 (very high), it is a relative measure of the amount of air pollution. They are based on 3 pollutants of major concern in Europe: $PM10$, $NO2$, $O3$ and will be able to take into account to 3 additional pollutants (CO, $PM2.5$ and $SO2$) where data are also available.

The calculation of the index is based on a review of a number of existing air quality indices, and it reflects EU alert threshold levels or daily limit values as much as possible. In order to make cities more comparable, independent of the nature of their monitoring network two situations are defined:

- Background, representing the general situation of the given agglomeration (based on urban background monitoring sites),

- Roadside, being representative of city streets with a lot of traffic, (based on roadside monitoring stations)

The indices values are updated hourly (for those cities that supply hourly data) and yesterdays daily indices are presented.

Common Air Quality Index Legend:

Pollution	Index Value
Very low	0/25
Low	25/50
Medium	50/75
High	75/100
Very high	>100

The Common Annual Air Quality Index

The common annual air quality index provides a general overview of the air quality situation in a given city all the year through and regarding to the European norms.

It is also calculated both for background and traffic conditions but its principle of calculation is different from the hourly and daily indices. It is presented as a distance to a target index, this target being derived from the EU directives (annual air quality standards and objectives):

- If the index is higher than 1: for one or more pollutants the limit values are not met.

- If the index is below 1: on average the limit values are met.

The annual index is aimed at better taking into account long term exposure to air pollution based on distance to the target set by the EU annual norms, those norms being linked most of the time to recommendations and health protection set up by World Health Organisation.

United States

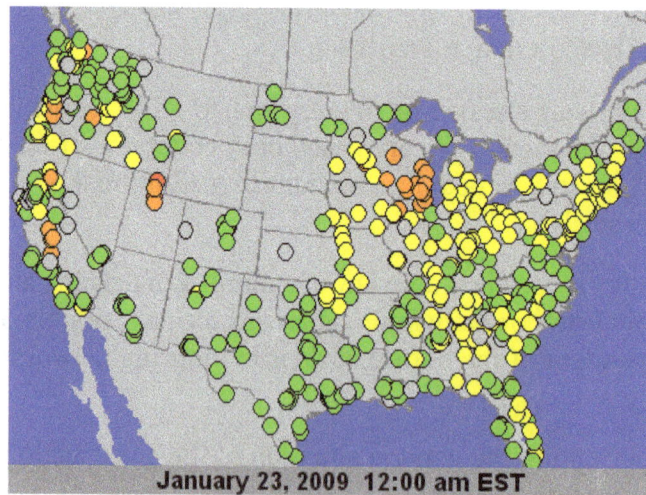

January 23, 2009 12:00 am EST

PM$_{2.5}$ 24-Hour AQI Loop, Courtesy US EPA

The United States Environmental Protection Agency (EPA) has developed an Air Quality Index that is used to report air quality. This AQI is divided into six categories indicating increasing levels of health concern. An AQI value over 300 represents hazardous air quality and below 50 the air quality is good.

Air Quality Index (AQI) Values	Levels of Health Concern	Colors
0 to 50	Good	Green
51 to 100	Moderate	Yellow
101 to 150	Unhealthy for Sensitive Groups	Orange
151 to 200	Unhealthy	Red
201 to 300	Very Unhealthy	Purple
301 to 500	Hazardous	Maroon

The AQI is based on the five "criteria" pollutants regulated under the Clean Air Act: ground-level ozone, particulate matter, carbon monoxide, sulfur dioxide, and nitrogen dioxide. The EPA has established National Ambient Air Quality Standards (NAAQS) for each of these pollutants in order to protect public health. An AQI value of 100 generally corresponds to the level of the NAAQS for the pollutant. The Clean Air Act (USA) (1990) requires EPA to review its National Ambient Air Quality Standards every five years to reflect evolving health effects information. The Air Quality Index is adjusted periodically to reflect these changes.

Computing the AQI

The air quality index is a piecewise linear function of the pollutant concentration. At the boundary between AQI categories, there is a discontinuous jump of one AQI unit. To convert from concentration to AQI this equation is used:

$$I = \frac{I_{high} - I_{low}}{C_{high} - C_{low}}(C - C_{low}) + I_{low}$$

where:

I = the (Air Quality) index,

C = the pollutant concentration,

C_{low} = the concentration breakpoint that is $\leq C$,

C_{high} = the concentration breakpoint that is $\geq C$,

I_{low} = the index breakpoint corresponding to C_{low},

I_{high} = the index breakpoint corresponding to C_{high}..

EPA's table of breakpoints is:

O_3 (ppb)	O_3 (ppb)	$PM_{2.5}$ (µg/m³)	PM_{10} (µg/m³)	CO (ppm)	SO_2 (ppb)	NO_2 (ppb)	AQI	AQI
C_{low} - C_{high} (avg)	C_{low} - C_{high} (avg)	C_{low} - C_{high} (avg)	C_{low} - C_{high} (avg)	C_{low} - C_{high} (avg)	C_{low} - C_{high} (avg)	C_{low} - C_{high} (avg)	I_{low} - I_{high}	Category
0-54 (8-hr)	-	0.0-12.0 (24-hr)	0-54 (24-hr)	0.0-4.4 (8-hr)	0-35 (1-hr)	0-53 (1-hr)	0-50	Good
55-70 (8-hr)	-	12.1-35.4 (24-hr)	55-154 (24-hr)	4.5-9.4 (8-hr)	36-75 (1-hr)	54-100 (1-hr)	51-100	Moderate
71-85 (8-hr)	125-164 (1-hr)	35.5-55.4 (24-hr)	155-254 (24-hr)	9.5-12.4 (8-hr)	76-185 (1-hr)	101-360 (1-hr)	101-150	Unhealthy for Sensitive Groups
86-105 (8-hr)	165-204 (1-hr)	55.5-150.4 (24-hr)	255-354 (24-hr)	12.5-15.4 (8-hr)	186-304 (1-hr)	361-649 (1-hr)	151-200	Unhealthy
106-200 (8-hr)	205-404 (1-hr)	150.5-250.4 (24-hr)	355-424 (24-hr)	15.5-30.4 (8-hr)	305-604 (24-hr)	650-1249 (1-hr)	201-300	Very Unhealthy
-	405-504 (1-hr)	250.5-350.4 (24-hr)	425-504 (24-hr)	30.5-40.4 (8-hr)	605-804 (24-hr)	1250-1649 (1-hr)	301-400	Hazardous
-	505-604 (1-hr)	350.5-500.4 (24-hr)	505-604 (24-hr)	40.5-50.4 (8-hr)	805-1004 (24-hr)	1650-2049 (1-hr)	401-500	Hazardous

Suppose a monitor records a 24-hour average fine particle ($PM_{2.5}$) concentration of 12.0 micrograms per cubic meter. The equation above results in an AQI of:

corresponding to air quality in the "Good" range. To convert an air pollutant concentration to an AQI, EPA has developed a calculator.

If multiple pollutants are measured at a monitoring site, then the largest or "dominant" AQI value is reported for the location. The ozone AQI between 100 and 300 is computed by selecting the larger of the AQI calculated with a 1-hour ozone value and the AQI computed with the 8-hour ozone value.

8-hour ozone averages do not define AQI values greater than 300; AQI values of 301 or greater are calculated with 1-hour ozone concentrations. 1-hour SO_2 values do not define higher AQI values greater than 200. AQI values of 201 or greater are calculated with 24-hour SO_2 concentrations.

Real time monitoring data from continuous monitors are typically available as 1-hour averages. However, computation of the AQI for some pollutants requires averaging over multiple hours of data. (For example, calculation of the ozone AQI requires computation of an 8-hour average and computation of the $PM_{2.5}$ or PM_{10} AQI requires a 24-hour average.) To accurately reflect the current air quality, the multi-hour average used for the AQI computation should be centered on the current time, but as concentrations of future hours are unknown and are difficult to estimate accurately, EPA uses surrogate concentrations to estimate these multi-hour averages. For reporting the $PM_{2.5}$, PM_{10} and ozone air quality indices, this surrogate concentration is called the NowCast. The Nowcast is a particular type of weighted average that provides more weight to the most recent air quality data when air pollution levels are changing.

Public Availability of the AQI

Real time monitoring data and forecasts of air quality that are color-coded in terms of the air quality index are available from EPA's AirNow web site. Historical air monitoring data including AQI charts and maps are available at EPA's AirData website.

History of the AQI

The AQI made its debut in 1968, when the National Air Pollution Control Administration undertook an initiative to develop an air quality index and to apply the methodology to Metropolitan Statistical Areas. The impetus was to draw public attention to the issue of air pollution and indirectly push responsible local public officials to take action to control sources of pollution and enhance air quality within their jurisdictions.

Jack Fensterstock, the head of the National Inventory of Air Pollution Emissions and Control Branch, was tasked to lead the development of the methodology and to compile the air quality and emissions data necessary to test and calibrate resultant indices.

The initial iteration of the air quality index used standardized ambient pollutant concentrations to yield individual pollutant indices. These indices were then weighted and summed to form a single total air quality index. The overall methodology could use concentrations that are taken from ambient monitoring data or are predicted by means of a diffusion model. The concentrations were then converted into a standard statistical distribution with a preset mean and standard deviation. The resultant individual pollutant indices are assumed to be equally weighted, although values other than unity can be used. Likewise, the index can incorporate any number of pollutants although it was only used to combine SOx, CO, and TSP because of a lack of available data for other pollutants.

While the methodology was designed to be robust, the practical application for all metropolitan areas proved to be inconsistent due to the paucity of ambient air quality monitoring data, lack of agreement on weighting factors, and non-uniformity of air quality standards across geographical and political boundaries. Despite these issues, the publication of lists ranking metropolitan areas achieved the public policy objectives and led to the future development of improved indices and their routine application.

Early Life Stage Test

An early life stage (ELS) test is a chronic toxicity test using sensitive early life stages like embryos or larvae to predict the effects of toxicants on organisms. ELS tests were developed to be quicker and more cost-efficient than full life-cycle tests, taking on average 1-5 months to complete compared to 6-12 months for a life-cycle test. They are commonly used in aquatic toxicology, particularly with fish. Growth and survival are the typically measured endpoints, for which a Maximum Acceptable Toxicant Concentration (MATC) can be estimated. ELS tests allow for the testing of fish species that otherwise could not be studied due to length of life, spawning requirements, or size. ELS tests are used as part of environmental risk assessments by regulatory agencies including the

U.S. Environmental Protection Agency (EPA) and Environment Canada, as well as the Organisation for Economic Co-operation and Development (OECD).

Development

ELS tests were adapted from full life-cycle toxicity tests, chronic tests that expose an organism to a contaminant for its entire life-cycle. These are widely considered to be the best tests for estimating long-term "safe" concentrations of toxicants in aquatic organisms. The first full life-cycle tests on fish were developed for the fathead minnow (*Pimephales promelas*), and later for bluegill (*Lepomis macrochirus*), brook trout (*Salvelinus fontinalis*), flagfish (*Jordanella floridae*), and sheepshead minnow (*Cyprinodon variegatus*). While useful, full life-cycle tests require a high number of test organisms and extensive exposure time in the lab, especially for vertebrates. Typically, life-cycle tests take 6-12 months for fathead minnow and 30 months for brook trout.

Following the passage of the Toxic Substances Control Act (TSCA) in the United States in 1976, there was an increased need for quicker, more efficient vertebrate toxicity tests. The EPA was now required to assess the environmental effects of new chemicals before they could be commercially produced. Less costly and time-intensive tests were needed to evaluate a multitude of new chemicals. Researchers began developing toxicity tests that focused on early life stages, since these have been shown to be more sensitive to environmental stressors than later life stages. Many critical events occur in a short period of time in the early stages of development. If a stressor disrupts developmental events (including their timing), it could result in adverse effects that reduce the organism's chances of survival. Meta-analysis has found that early life-cycle portions of full life-cycle tests usually estimate an MATC within a factor of 2 of full life-cycle estimates in saltwater and freshwater fish. In 83% of 72 tests, the ELS portion resulted in the same MATC as the full life-cycle estimate, and the remaining 17% were within a factor of 2.

Limitations

There remain some limitations with early life stage toxicity testing. Although ELS tests are quicker and more cost-efficient than full life cycle tests, they remain resource- and time-intensive. Fish early life stage (FELS) tests require hundreds of fish and 1 to 5 months to complete. Other issues include the lack of mechanistic information, differing sensitivities between species, and insensitivity to parental exposure. ELS tests don't provide information on the toxicant's mechanism of action. Sensitivity to specific toxicants varies with species, so the most sensitive or most important species should be tested in each case. ELS tests appear to be insensitive to parental exposure, and MATCs are generally the same for embryos of both exposed and unexposed parents. This could be due to the mode of action of the toxicant or the variability and insensitivity of ELS test design. Additionally, growth response has been found to be an insensitive endpoint in ELS tests with fish, having little bearing on the estimation of an MATC. Growth response could be omitted to reduce the duration and cost of screening tests.

Methodology

In a typical early life stage toxicity test, a flow-through dilutor system administers different concentrations of a toxicant to different test chambers. At least five different concentrations of a toxicant are tested, plus controls, with at least two exposure chambers for each treatment. The length of the exposure depends on the test species. For example, fathead minnow tests are 1-2 months long, while brook trout tests are around 5 months long. Growth and survival are the typical endpoints, for which an MATC can be found.

Standard methods for ELS tests have been established by the OECD, ASTM International, the EPA, and Environment Canada.

Regulatory Uses

- The FELS guideline of the OECD Guidelines for the Testing of Chemicals is the primary test used internationally to estimate chronic fish toxicity.

- FELS tests are part of a suite of sublethal toxicity tests for effluent used by Environment Canada in environmental effects monitoring.

- A FELS test is required or recommended by the US EPA for testing and monitoring chemicals released into aquatic systems.

Current Developments

An extended ELS test has been examined as a potential surrogate for a fish full life-cycle test to detect weak environmental estrogens. Endocrine active chemicals (EACs) are ubiquitous in the environment, prompting the need for better screening assays to predict their effects, especially in aquatic species. Slightly longer ELS tests could be used instead of full life-cycle tests, taking into account sensitive windows of exposure like sexual differentiation and early gonadal development. Extended ELS tests have proven successful in detecting the effects of weak estrogens in fathead minnows.

Additionally, adverse outcome pathways (AOPs) are being used to develop an alternative to FELS testing. Industry and regulatory agencies are increasingly interested in an animal-free, cost-efficient surrogate. Researchers are developing FELS-related AOPs to create a high-throughput, less costly screening strategy for toxicants that takes the mechanism of action into account.

Modes of Toxic Action

A mode of toxic action is a common set of physiological and behavioral signs that characterize a type of adverse biological response. A mode of action should not be confused with mechanism of action, which refer to the biochemical processes underlying a given mode of action. Modes of toxic action are important, widely used tools in ecotoxicology and aquatic toxicology because they classify toxicants or pollutants according to their type of toxic action. There are two major types of modes of toxic action: non-specific acting toxicants and specific acting toxicants. Non-specific acting toxicants are those that produce narcosis, while specific acting toxicants are those that are non-narcotic and that produce a specific action at a specific target site.

Types

Non-specific

Non-specific acting modes of toxic action result in narcosis; therefore, narcosis is a mode of toxic action. Narcosis is defined as a generalized depression in biological activity due to the presence of toxicant molecules in the organism. The target site and mechanism of toxic action through which narcosis affects organisms are still unclear, but there are hypotheses that support that it occurs through alterations in the cell membranes at specific sites of the membranes, such as the lipid layers or the proteins bound to the membranes. Even though continuous exposure to a narcotic toxicant can produce death, if the exposure to the toxicant is stopped, narcosis can be reversible.

Specific

Toxicants that at low concentrations modify or inhibit some biological process by binding at a specific site or molecule have a specific acting mode of toxic action. However, at high enough concentrations, toxicants with specific acting modes of toxic actions can produce narcosis that may or may not be reversible. Nevertheless, the specific action of the toxicant is always shown first because it requires lower concentrations.

There are several specific acting modes of toxic action:

- *Uncouplers of oxidative phosphorylation.* Involves toxicants that uncouple the two processes that occur in oxidative phosphorylation: electron transfer and adenosine triphosphate (ATP) production.

- *Acetylcholinesterase (AChE) inhibitors.* AChE is an enzyme associated with nerve synapses that it's designed to regulate nerve impulses by breaking down the neurotransmitter Acetylcholine (ACh). When toxicants bind to AChE, they inhibit the breakdown of ACh. This results in continued nerve impulses across the synapses, which eventually cause nerve system damage. Examples of AChE inhibitors are organophosphates and carbamates, which are components found in pesticides.

- *Irritants.* These are chemicals that cause an inflammatory effect on living tissue by chemical action at the site of contact. The resulting effect of irritants is an increase in the volume of cells due to a change in size (hypertrophy) or an increase in the number of cells (hyperplasia). Examples of irritants are benzaldehyde, acrolein, zinc sulphate and chlorine.

- *Central nervous system (CNS) seizure agents.* CNS seizure agents inhibit cellular signaling by acting as receptor antagonists. They result in the inhibition of biological responses. Examples of CNS seizure agents are organochlorine pesticides.

- *Respiratory blockers.* These are toxicants that affect respiration by interfering with the electron transport chain in the mitochondria. Examples of respiratory blockers are rotenone and cyanide.

Determination

The pioneer work of identifying the major categories of modes of toxic action was conducted by investigators from the U.S. Environmental Protection Agency (EPA) at the Duluth Laboratory using fish, reason why they named the categories as *Fish Acute Toxicity Syndromes* (FATS). They proposed the FATS by assessing the behavioral and physiological responses of the fish when subjected to toxicity tests, such as locomotive activities, body color, ventilation patterns, cough rate, heart rate, and others.

It has been proposed that modes of toxic action could be estimated by developing a data set of critical body residues (CBR). The CBR is the whole-body concentration of a chemical that is associated with a given adverse biological response and it is estimated using a partition coefficient and a bioconcentration factor. The whole-body residues are reasonable first approximations of the amount of chemical present at the toxic action site(s). Because different modes of toxic action generally appear to be associated with different ranges of body residues, modes of toxic action can then be separated into categories. However, it is unlikely that every chemical has the same mode of toxic action in every organism, so this variability should be considered. The effects of mixture toxicity should be considered as well, even though mixture toxicity it's generally additive, chemicals with more than one mode of toxic action may contribute to toxicity.

Modeling has become a common used tool to predict modes of toxic action in the last decade. The models are based in Quantitative Structure-Activity Relationships (QSARs), which are mathematical models that relate the biological activity of molecules to their chemical structures and corresponding chemical and physicochemical properties. QSARs can then predict modes of toxic action of unknown compounds by comparing its characteristic toxicity profile and chemical structure to reference compounds with known toxicity profiles and chemical structures. Russom and colleagues were one of the first group of researchers being able to classify modes of toxic action with the use of QSARs; they classified 600 chemicals as narcotics. Even though QSARs are a useful tool for predicting modes of toxic action, chemicals having multiple modes of toxic action can obscure QSAR analyses. Therefore, these models are continuously being developed.

Applications

Environmental Risk Assessment

The objective of environmental risk assessment is to protect the environment from adverse effects. Researchers are further developing QSAR models with the ultimate goal providing a clear insight about a mode of toxic action, but also about what the actual target site is, the concentration of the chemical at this target site, and the interaction occurring at the target site, as well as to predict the modes of toxic action in mixtures. Information on the mode of toxic action is crucial not only in understanding joint toxic effects and potential interactions between chemicals in mixtures, but also for developing assays for the evaluation of complex mixtures in the field.

Regulation

The combination of behavioral and physiological responses, CBR estimates, and chemical fate and bioaccumulation QSAR models can be a powerful regulatory tool to address pollution and toxicity in areas where effluents are discharged.

Exposure Assessment

Exposure assessment is a branch of environmental science and occupational hygiene that focuses on the processes that take place at the interface between the environment containing the contaminant(s) of interest and the organism(s) being considered. These are the final steps in the path to release an environmental contaminant, through transport to its effect in a biological system. It tries to measure how much of a contaminant can be absorbed by an exposed target organism, in what form, at what rate and how much of the absorbed amount is actually available to produce a biological effect. Although the same general concepts apply to other organisms, the overwhelming majority of applications of exposure assessment are concerned with human health, making it an important tool in public health.

Definition

Exposure assessment is the process of estimating or measuring the magnitude, frequency and duration of exposure to an agent, along with the number and characteristics of the population exposed. Ideally, it describes the sources, pathways, routes, and the uncertainties in the assessment.

Exposure analysis is the science that describes how an individual or population comes in contact with a contaminant, including quantification of the amount of contact across space and time. 'Exposure assessment' and 'exposure analysis' are often used as synonyms in many practical contexts. Risk is a function of exposure and hazard. For example, even for an extremely toxic (high hazard) substance, the risk of an adverse outcome is unlikely if exposures are near zero. Conversely, a moderately toxic substance may present substantial risk if an individual or a population is highly exposed.

Applications

Quantitative measures of exposure are used: in risk assessment, together with inputs from toxicology, to determine risk from substances released to the environment, to establish protective standards, in epidemiology, to distinguish between exposed and control groups, and to protect workers from occupational hazards.

Receptor-based Approach

The receptor-based approach is used in exposure science. It starts by looking at different contaminants and concentration that reach people. An exposure analyst can use direct or indirect measurements to determine if a person has been in contact with a specific contaminant.Once a contaminant has been proved to reach people, exposure analysts work backwards to determine its source. After the identification of the source, it is important to find out the most efficient way to reduce adverse health effects. If the contaminant reaches a person, it is very hard to reduce the associated adverse effects. Therefore, it is very important to reduce exposure in order to diminish the risk of adverse health effects. It is highly important to use both regulatory and non-regulatory approaches in order to decrease people's exposure to contaminants. In many cases, it is better to change people's activities in order to reduce their exposures rather than regulating a source of contaminants. The receptor-based approach can be opposed to the source-based approach. This approach begins by looking at different sources of contaminants such as industries and power plants. Then, it is important to find out if the contaminant of interest has reached a receptor (usually humans). With this approach, it is very hard to prove that a pollutant from a source has reached a target.

Exposure

In this context *exposure* is defined as the contact between an agent and a target. Contact takes place at an exposure surface over an exposure period. Mathematically, exposure is defined as

$$E = \int_{t_1}^{t_2} C(t)dt$$

where E is exposure, $C(t)$ is a concentration that varies with time between the beginning and end of exposure. It has dimensions of mass times time divided by volume. This quantity is related to the potential dose of contaminant by multiplying it by the relevant contact rate, such as breathing rate, food intake rate etc. The contact rate itself may be a function of time.

Routes of Exposure

Contact between a contaminant and an organism can occur through any route. The possible routes of exposure are: inhalation, if the contaminant is present in the air, ingestion, through food, drinking or hand-to-mouth behavior, and dermal absorption, if the contaminant can be absorbed through the skin.

Exposure to a contaminant can and does occur through multiple routes, simultaneously or at different times. In many cases the main route of exposure is not obvious and needs to be investigated carefully. For example, exposure to byproducts of water chlorination can obviously occur by drinking, but also through the skin, while swimming or washing, and even through inhalation from droplets aerosolized during a shower. The relative proportion of exposure from these different routes cannot be determined *a priori*. Therefore, the equation in the previous section is correct in a strict mathematical sense, but it is a gross oversimplification of actual exposures, which are the sum of the integrals of all activities in all microenvironments. For example, the equation would have to be calculated with the specific concentration of a compound in the air in the room during the time interval. Similarly, the concentration in the ambient air would apply to the time that the person spends outdoors, whereas the concentration in the food that the person ingests would be added. The concentration integrals via all routes would be added for the exposure duration, e.g. hourly, daily or annually as

$$E = sum(\int_{t_1}^{t_2} C(t)dt ... \int_{t_y}^{t_z} C(t)dt)$$

where y is the initial time and z the ending time of last in the series of time periods spent in each microenvironment over the exposure duration.

Measurement of Exposure

To quantify the exposure of particular individuals or populations two approaches are used, primarily based on practical considerations:

Direct Approach

The direct approach measures the exposures to pollutants by monitoring the pollutant concentrations reaching the respondents. The pollutant concentrations are directly monitored on or within the person through point of contact, biological monitoring, or biomarkers. The point of contact approach indicates the total concentration reaching the host, while biological monitoring and the use of biomarkers infer the dosage of the pollutant through the determination of the body burden. The respondents often record their daily activities and locations during the measurement of the pollutants to identify the potential sources, microenvironments, or human activities contributing the pollutant exposure. An advantage of the direct approach is that the exposures through multiple media (air, soil, water, food, etc.) are accounted for through one study technique. The disadvantages include the invasive nature of the data collection and associated costs.

Point of contact is continuous measure of the contaminant reaching the target through all routes.

Biological monitoring is another approach to measuring exposure. This measures the amount of a pollutant within the body in various tissue media such adipose tissue, bone, or urine. Biological monitoring measures the body burden of a pollutant but not the source from whence it came. The substance measured may be either the contaminant itself or a biomarker which is specific to and indicative of an exposure to the contaminant.

Biomarkers of exposure assessment is a measure of the contaminant or other proportionally related variable in the body.

Air sampling measures the contaminant in the air as concentration units of ppmv (parts per million by volume), mg/m^3 (milligrams per cubic meter) or other mass per unit volume of air. Samples can be worn by workers or researchers to estimate concentrations found in the breathing zone (personal) or samples collected in general areas that can be used to estimate human exposure by integrating time and activity patterns. Validated and semi-validated air sampling methods are published by NIOSH, OSHA, ISO and other bodies.

Surface or dermal sampling measures of the contaminant on touchable surfaces or on skin. Concentrations are typically reported in mass per unit surface area such as $mg/100 \ cm^2$. Validated and semi-validated air sampling methods are published by NIOSH, US EPA, OSHA, ISO and other bodies.

Indirect Approach

The indirect approach measures the pollutant concentrations in various locations or during specific human activities to predict the exposure distributions within a population. The indirect approach focuses on the pollutant concentrations within microenvironments or activities rather than the concentrations directly reaching the respondents. The measured concentrations are correlated to large-scale activity pattern data, such as the National Human Activity Pattern Survey (NHAPS), to determine the predicted exposure by multiplying the pollutant concentrations by the time spent in each microenvironment or activity for by multiplying the pollutant concentrations b the contact rate with each media. The indirect approach or exposure modeling determines the estimated exposure distributions within a population rather than the direct exposure an individual has experienced. The advantage is that process is minimally invasive to the population and is associated with lower costs than the direct approach. A disadvantage of the approach is that the results were determined independently of any actual exposures, so the exposure distribution is open to errors from any inaccuracies in the assumptions made during the study, the time-activity data, or the measured pollutant concentrations.

In general, direct methods tend to be more accurate but more costly in terms of resources and demands placed on the subject being measured and may not always be feasible, especially for a population exposure study.

Examples of direct methods include air sampling though a personal portable pump, split food samples, hand rinses, breath samples or blood samples. Examples of indirect methods include environmental water, air, dust, soil or consumer product sampling coupled with information such as activity/location diaries. Mathematical exposure models may also be used to explore hypothetical situations of exposure.

Exposure Factors

Especially when determining the exposure of a population rather than individuals, indirect methods can often make use of relevant statistics about the activities that can lead to an exposure. These statistics are called *exposure factors*. They are generally drawn from the scientific literature or governmental statistics. For example they may report informations such as amount of different food eaten by specific populations, divided by location or age, breathing rates, time spent for different modes of commuting, showering or vacuuming, as well as information on types of residences. Such information can be combined with contaminant concentrations from *ad-hoc* studies or monitoring network to produce estimates of the exposure in the population of interest. These are especially useful in establishing protective standards.

Exposure factor values can be used to obtain a range of exposure estimates such as average, high-end and bounding estimates. For example, to calculate the lifetime average daily dose one would use the equation below:

$LADD = (ContaminantConcentration)(IntakeRate)$

$(ExposureDuration)/(BodyWeight)(AverageLifetime)$

All of the variables in the above equation, with the exception of contaminant concentration, are considered exposure factors. Each of the exposure factors involves humans, either in terms of their characteristics (e.g., body weight) or behaviors (e.g., amount of time spent in a specific location, which affects exposure duration). These characteristics and behaviors can carry a great deal of variability and uncertainty. In the case of lifetime average daily dose, variability pertains to the distribution and range of LADDs amongst individuals in the population. The uncertainty, on the other hand, refers to exposure analyst's lack of knowledge of the standard deviation, mean, and general shape when dealing with calculating LADD.

The U.S. Environmental Protection Agency's *Exposure Factors Handbook* provides solutions when confronting variability and reducing uncertainty. The general points are summarized below:

Four Strategies for Confronting Variability	Examples
Disaggregate the variability	Develop distribution of body weight for subgroup
Ignore the variability	Assume all adults weigh 65 kg
Use a maximum or minimum value	Choose a high-end value for weight distribution
Use the average value	Use the mean body weight for all adults

Analyzing Uncertainty	Description
Classical statistical methods (descriptive statistics and inferential statistics)	Estimating the population exposure distribution directly, based on measured values from a representative sample
Sensitivity analysis	Changing one input variable at a time while leaving others constant, to examine effect on output
Propagation of uncertainty	Examining how uncertainty in individual parameters affects the overall uncertainty of the exposure assessment
Probabilistic analysis	Varying each of the input variables over various values of their respective probability distributions(i.e. Monte Carlo integration)

Defining Acceptable Exposure for Occupational Environments

Hierarchies for Effective and Efficient Protection of Workers & Communities

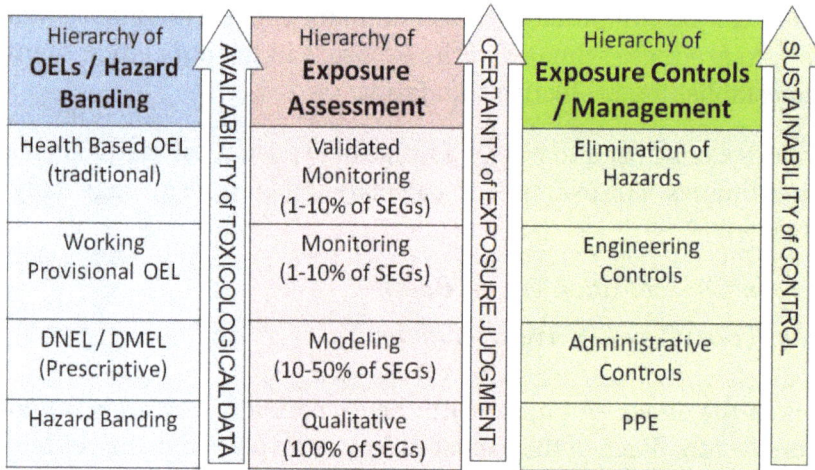

Simple representation of exposure risk assessment and management hierarchy based on available information

Occupational exposure limits are based on available toxicology and epidemiology data to protect nearly all workers over a working lifetime. Exposure assessments in occupational settings are most often performed by occupational/industrial hygiene (OH/IH) professionals who gather "basic characterization" consisting of all relevant information and data related to workers, agents of concern, materials, equipment and available exposure controls. The exposure assessment is initiated by selecting the appropriate exposure limit averaging time and "decision statistic" for the agent. Typically the statistic for deciding acceptable exposure is chosen to be the majority (90%, 95% or 99%) of all exposures to be below the selected occupational exposure limit. For retrospective exposure assessments performed in occupational environments, the "decision statistic" is typically a central tendency such as the arithmetic mean or geometric mean or median for each worker or group of workers. Methods for performing occupational exposure assessments can be found in "A Strategy for Assessing and Managing Occupational Exposures".

Exposure assessment is a continuous process that is updated as new information and data becomes available.

Bioanalysis

Bioanalysis is a sub-discipline of analytical chemistry covering the quantitative measurement of xenobiotics (drugs and their metabolites, and biological molecules in unnatural locations or concentrations) and biotics (macromolecules, proteins, DNA, large molecule drugs, metabolites) in biological systems.

Modern Bioanalytical Chemistry

Many scientific endeavors are dependent upon accurate quantification of drugs and endogenous substances in biological samples; the focus of bioanalysis in the pharmaceutical industry is to provide a quantitative measure of the active drug and/or its metabolite(s) for the purpose of pharmacokinetics, toxicokinetics, bioequivalence and exposure–response (pharmacokinetics/pharmacodynamics studies). Bioanalysis also applies to drugs used for illicit purposes, forensic investigations, anti-doping testing in sports, and environmental concerns.

Bioanalysis was traditionally thought of in terms of measuring small molecule drugs. However, the past twenty years has seen an increase in biopharmaceuticals (e.g. proteins and peptides), which have been developed to address many of the same diseases as small molecules. These larger biomolecules have presented their own unique challenges to quantification.

History

The first studies measuring drugs in biological fluids were carried out to determine possible overdosing as part of the new science of forensic medicine/toxicology.

Initially, nonspecific assays were applied to measuring drugs in biological fluids. These were unable to discriminate between the drug and its metabolites; for example, aspirin (circa 1900) and sulfonamides (developed in the 1930s) were quantified by the use of colorimetric assays. Antibiotics were quantified by their ability to inhibit bacterial growth. The 1930s also saw the rise of pharmacokinetics, and as such the desire for more specific assays. Modern drugs are more potent, which has required more sensitive bioanalytical assays to accurately and reliably determine these drugs at lower concentrations. This has driven improvements in technology and analytical methods.

Bioanalytical Techniques

Some techniques commonly used in bioanalytical studies include:

- Hyphenated techniques

    o LC–MS (liquid chromatography–mass spectrometry)

    o GC–MS (gas chromatography–mass spectrometry)

    o LC–DAD (liquid chromatography–diode array detection)

    o CE–MS (capillary electrophoresis–mass spectrometry)

- Chromatographic methods

    o HPLC (high performance liquid chromatography)

    o GC (gas chromatography)

    o UPLC (ultra performance liquid chromatography)

    o Supercritical fluid chromatography

- Electrophoresis

- Ligand binding assays

    o Dual polarisation interferometry

    o ELISA (Enzyme-linked immunosorbent assay)

    o MIA (magnetic immunoassay)

    o RIA (radioimmunoassay)

- Mass spectrometry

- Nuclear magnetic resonance

The most frequently used techniques are: liquid chromatography coupled with tandem mass spectrometry (LC–MS/MS) for 'small' molecules and enzyme-linked immunosorbent assay (ELISA) for macromolecules

Sample Preparation and Extraction

The bioanalyst deals with complex biological samples containing the analyte alongside a diverse range of chemicals that can have an adverse impact on the accurate and precise quantification of the analyte. As such, a wide range of techniques are applied to extract the analyte from its matrix. These include:

- Protein precipitation

- Liquid–liquid extraction

- Solid phase extraction

Bioanalytical laboratories often deal with large numbers of samples, for example resulting from clinical trials. As such, automated sample preparation methods and liquid-handling robots are commonly employed to increase efficiency and reduce costs.

Bioanalytical Organisations

There are several national and international bioanalytical organisations active throughout the world. Often they are part of a bigger organisation, e.g. Bioanalytical Focus Group and Ligand Binding Assay Bioanalytical Focus Group, which are both within the American Association of Pharmaceutical Scientists (AAPS) and FABIAN, a working group of the Analytical Chemistry Section of the Royal Netherlands Chemical Society. The European Bioanalysis Forum (EBF), on the other hand, is independent of any larger society or association.

Toxic Equivalency Factor

Toxic equivalency factor (TEF) expresses the toxicity of dioxins, furans and PCBs in terms of the most toxic form of dioxin, 2,3,7,8-TCDD. The toxicity of the individual congeners may vary by orders of magnitude.

With the TEFs, the toxicity of a mixture of dioxins and dioxin-like compounds can be expressed in a single number - the toxic equivalency (TEQ). It is a single figure resulting from the product of the concentration and individual TEF values of each congener.

The TEF/TEQ concept has been developed to facilitate risk assessment and regulatory control. While the initial and current set of TEFs only apply to dioxins and dioxin-like chemicals (DLCs), the concept can theoretically be applied to any group of chemicals satisfying the extensive similarity criteria used with dioxins, primarily that the main mechanism of action is shared across the group. Thus far, only the DLCs have had such a high degree of evidence of toxicological similarity.

There have been several systems over the years in operation, such as the International Toxic Equivalents for dioxins and furans only, represented as $I\text{-}TEQ_{DF}$, as well as several country-specific TEFs. The present World Health Organisation scheme, represented as $WHO\text{-}TEQ_{DFP}$, which includes PCBs is now universally accepted.

Chemical Mixtures and Additivity

Humans and wildlife are rarely exposed to solitary contaminants, but rather to complex mixtures of potentially harmful compounds. Dioxins and DLCs are no exception. This is important to consider when assessing toxicity because the effects of chemicals in a mixture are often different from when acting alone. These differences can take place on the chemical level, where the properties of the compounds themselves change due to the interaction, creating a new dose at the target tissue and a quantitatively different effect. They may also act together (simple similar action) or independently on the organism at the receptor during uptake, when transported throughout the body, or during metabolism, to produce a joint effect. Joint effects are described as being additive (using dose, response/risk, or measured effect), synergistic, or antagonistic. A dose-additive response occurs when the mixture effect is determined by the sum of the component chemical doses, each weighted by its relative toxic potency. A risk-additive response occurs when the mixture response is the sum of component risks, based on the probability law of independent events. An effect-additive mixture response occurs when the combined effect of exposure a chemical mixture is *equal to* the sums of the separate component chemical effects, e.g., incremental changes in relative liver weight. Synergism occurs when the combined effect of chemicals together is *greater than* the additivity prediction based on their separate effects. Antagonism describes where the combined effect is *less than* the additive prediction. Clearly it is important to identify which kind of additivity is being used. These effects reflect the underlying modes of action and mechanisms of toxicity of the chemicals.

Additivity is an important concept here because the TEF method operates under the assumption that the assessed contaminants are dose-additive in mixtures. Because dioxins and DLCs act similarly at the AhR, their individual quantities in a mixture can be added together as proportional values, i.e. TEQs, to assess the total potency. This notion is fairly well supported by research. Some interactions have been observed and some uncertainties remain, including application to other than oral intake.

TEF

Exposure to environmental media containing 2,3,7,8-TCDD and other dioxins and dioxin-like compounds can be harmful to humans as well as to wildlife. These chemicals are resistant to metabolism and biomagnify up the food chain. Toxic and biological effects of these compounds are mediated through the aryl hydrocarbon receptor (AhR). Oftentimes results of human activity leads to instances of these chemicals as mixtures of DLCs in the environment. The TEF approach has also been used to assess the toxicity of other chemicals including PAHs and xenoestrogens.

The TEF approach uses an underlying assumption of additivity associated with these chemicals that takes into account chemical structure and behavior. For each chemical the model uses comparative measures from individual toxicity assays, known as relative effect potency (REP), to assign a single scaling factor known as the TEF.

Toxic equivalency factor according to different schemes				
Congener	**BGA 1985**	**NATO (I-TEF) 1988**	**WHO 1998**	**WHO 2005**
2,3,7,8-Cl$_4$DD	1	1	1	1
1,2,3,7,8-Cl$_5$DD	0,1	0,5	1	1
2,3,7,8-subst. Cl$_6$DD	0,1	0,1	0,1	0,1
1,2,3,4,6,7,8-Cl$_7$DD	0,01	0,01	0,01	0,01
Cl$_8$DD	0,001	0,001	0,0001	0,0003
2,3,7,8-Cl$_4$DF	0,1	0,1	0,1	0,1
1,2,3,7,8-Cl$_5$DF	n.n.	n.n.	0,05	0,03
2,3,4,7,8-Cl$_5$DF	0,01	0,05	0,5	0,3
2,3,7,8-subst. Cl$_6$DF	0,01	0,01	0,01	0,01
2,3,7,8-subst. Cl$_7$DF	0,01	0,01	0,01	0,01
other Cl$_7$DF	0,001	0	0	0
Cl$_8$DF	0,001	0,001	0,0001	0,0003
other PCDD and PCDF	0,01	0	0	0

TCDD

2,3,7,8-tetrachlorodibenzo-p-dioxin (TCDD) is the reference chemical to which the toxicity of other dioxins and DLCs are compared. TCDD is the most toxic DLC known. Other dioxins and DLCs are assigned a scaling factor, or TEF, in comparison to TCDD. TCDD has a TEF of 1.0. Sometimes PCB126 is also used as a reference chemical, with a TEF of 0.1.

Determination of TEF

TEFs are determined using a database of REPs that meet WHO established criteria, using different biological models or endpoints and are considered estimates with an order of magnitude of uncertainty. The characteristics necessary for inclusion of a compound in the WHO's TEF approach include:

- Structural similarity to polychlorinated dibenzo-p-dioxins or polychlorinated dibenzofurans

- Capacity to bind to the aryl hydrocarbon receptor (AhR)

- Capacity to elicit AhR-mediated biochemical and toxic responses

- Persistence and accumulation in the food chain

All viable REPs for a chemical are compiled into a distribution, and the TEF is selected based on half order of magnitude increments on a logarithmic scale. The TEF is typically selected from the 75th percentile of the REP distribution in order to be protective of health.

In Vivo and *in Vitro* Studies

REP distributions are not weighted to give more importance to certain types of studies. Current focus of REPs is on *in vivo* studies rather than *in vitro*. This is because all types of *in vivo* studies (acute, subchronic, etc.) and different endpoints have been combined, and associated REP distributions are shown as a single box plot.

TEQ

Toxic Equivalents (TEQs) report the toxicity-weighted masses of mixtures of PCDDs, PCDFs, and PCBs. The reported value provides toxicity information about the mixture of chemicals and is more meaningful to toxicologists than reporting the total number of grams. To obtain TEQs the mass of each chemical in a mixture is multiplied by its TEF and is then summed with all other chemicals to report the total toxicity-weighted mass. TEQs are then used for risk characterization and management purposes, such as prioritizing areas of cleanup.

Calculation

The toxic equivalency of a mixture is defined by the sum of the concentrations of individual compounds (C_i) multiplied by their relative toxicity (TEF):

$$TEQ = \Sigma[C_i] \times TEF_i$$

Applications

Risk Assessment

Risk assessment is the process by which one estimates the probability of some adverse effect, such as that of a contaminant in the environment. Environmental risk assessments are conducted to help protect human health and the environment and are often used to assist in meeting regulations such as those stipulated by CERCLA in the United States. Risk assessments may take place retro-actively, i.e., when assessing the contamination hazard at a superfund site, or predictively, such as when planning waste discharges.

The complex nature of chemicals mixtures in the environment presents a challenge to risk assessment. The TEF approach was developed to help assess the toxicity of DLCs and other environmental contaminants with additive effects and is currently endorsed by the World Health Organization

Human Health

Human exposure to dioxins and DLCs is a cause for public and regulatory concern. Health concerns include endocrine, developmental, immune and carcinogenic effects. The route of exposure is primarily through the ingestion of animal products such as meat, dairy, fish, and human breast milk. However, humans are also exposed to high levels of "natural dioxins" in cooked foods and vegetables. The human diet accounts for over 95% of the total uptake of TEQ.

Risks in humans are typically calculated from known ingestion of contaminants or from blood or adipose tissue samples. However, human intake data is limited, and calculations from blood and tissue are not well supported. This presents a limitation to the TEF application in risk assessment to humans.

Fish and Wildlife

DLC exposure to wildlife results from various sources including the atmospheric deposition of emissions (e.g. waste incineration) over terrestrial and aquatic habitats and contamination from waste effluents. Contaminants then bioaccumulate up the food chain. The WHO has derived TEFs for fish, bird, and mammal species, however differences among taxa for some compounds are orders of magnitude apart. Compared to mammals, fish are less responsive to mono-ortho PCBs.

Limitations

The TEF approach DLC risk assessment operates under certain assumptions which attach varying degrees of uncertainty. These assumptions include:

- Individual compounds all act through the same biologic pathway

- Individual effects are dose-additive

- Dose-response curves are similarly shaped

- Individual compounds are similarly distributed throughout the body

TEFs are assumed to be equivalent for all effects, all exposure scenarios and all species, although this may not be the reality. The TEF method only accounts for toxicity effects related to the AhR mechanism - however, some DLC toxicity may be mediated through other processes. Dose-additivity may not be applicable to all DLCs and exposure scenarios, particularly those involving low doses. Interactions with other chemicals that may induce antagonistic effects are not considered and those may be species-specific. In terms of human health risk assessments, estimates of relative potency from animal studies are assumed to be predictive of toxicity in humans, although there are species-specific differences in the AhR. Nevertheless, *In vivo* mixture studies have shown that WHO 1998 TEF values predicted mixture toxicity within a factor of two or less A probabilistic approach may provide an advantage in the determination of TEF because it will better describe the level of uncertainty present in a TEF value

The use of TEF values to assess abiotic matrices such as soil, sediment, and water is problematic because TEF values are primarily calculated from oral intake studies.

History and Development

Dating back to the 1980s there is a long history of developing TEFs and how to use them. New research being conducted influences guiding criteria for assigning TEFs as the science progresses. The World Health Organization has held expert panels to reach a global consensus on how to assign TEFs in conjunction with new data. Each individual country recommends their own TEF values, typically endorsing the WHO global consensus TEFs.

Other Compounds for Potential Inclusion

Based on mechanistic considerations, PCB 37, PBDDs, PBDFs, PXCDDs, PXCDFs, PCNs, PBNs and PBBs can be included in the TEF concept. However, most of these compounds lack human exposure data. Thus, TEF values for these compounds are in the process of review

Blood Lead Level

Blood lead level (BLL), is a measure of lead in the blood. It is often measured in micrograms of lead per deciliter of blood (µg/dL) especially in the United States; 5 µg/dL is equivalent to 0.24 µmol/L (micromolar).

The Centers for Disease Control and Prevention (CDC) changed its view on blood lead levels in 2012 because of "a growing body of studies concluding that blood lead levels (BLLs) lower than 10 µg/dL harm children" with "irreversible" effects, and "since no safe blood lead level in children has been identified, a blood lead 'level of concern' cannot be used to define individuals in need of intervention". The new policy is to aim to reduce average blood lead levels in US children to as low a level as possible.

The CDC now publishes a "reference" blood lead level which they hope can decrease in coming years. The reference value is "based on the 97.5th percentile of the BLL distribution among children 1 –5 years old in the United States". It is currently 5 µg/dL. According to the CDC, in 2012, "approximately 450,000 children in the United States have BLLs higher than this reference value". There were more than 24 million US children under the age of 6 in 2014. If 2.5% are assumed to have blood lead levels higher than the reference amount, then there were approximately 600,000 US children with elevated blood lead levels in 2014. It is not a level deemed by the CDC as "safe". No level of lead in the blood of children is currently thought to be safe. The reference level is designed to be used as a policy tool. Parents, clinicians, communities, state and federal authorities and political leaders are expected to monitor blood lead test levels, aware that children testing higher than the reference level are testing higher than 97.5% of all US children. The CDC expects action to be taken when test levels are found to exceed the reference. As blood lead levels slowly decline in response to such action, the reference will also decline. CDC will recalculate a new reference every four years.

Pre-industrial human BLL measurements are estimated to have been 0.016 µg/dL, and this level increased markedly in the aftermath of the industrial revolution. Today, BLL measurements from remote human populations have ranged from 0.8 to 3.2 µg/dL. Children in populations adjacent to industrial centers in developing countries often have average BLL measurements above 25 µg/dL. The National Academies evaluated this issue in 1991 and confirmed that the blood lead level of the average person in the US was 300 - 500 times higher than that of preindustrial humans.

Lead is toxic and can cause neurological damage, especially among children, at any detectable level. High lead levels cause decreased vitamin D and haemoglobin synthesis, anemia, acute central nervous system disorders, and possibly death.

Historical Trends

Prior to the industrial revolution human BLL is estimated to have been far less than it is today. Bone lead measurements from two Native American populations living on the Pacific coast and the Colorado River between 1000-1300 A.D. show that BLLs would have been approximately 0.016 µg/dL. The World Health Organization and others interpret these measurements to be broadly representative of human preindustrial BLL.

Contemporary human BLLs in remote locations are estimated to be 0.8 and 3.2 µg/dL in the southern and northern hemispheres, respectively. Blood lead levels 50-1000 times higher than preindustrial levels are commonly measured in contemporary human populations around the world.

The baby on the left represents humans uncontaminated by industrial lead, the baby in the middle represents the typical American in 1980, and the baby on the right, people with overt clinical lead poisoning.

Created using data from Figure 1-1, "Measuring Lead Exposure in Infants, Children, and other Sensitive Populations (1993)" National Academies Press. The graphic published by the National Academies Press was itself adapted from NRC (National Research Council). 1980. "Lead in the Human Environment".

This is an adaptation of a graphic created by Clair Patterson. He originally developed techniques to measure tiny concentrations of lead in his quest to determine the age of Earth. When he discovered that preindustrial humans had far less lead in their bodies than all modern humans, he wrote: "It seems probable that persons polluted with amounts of lead that are at least 400 times higher than natural levels, and are nearly one-third to one-half that required to induce dysfunction, that their lives are being adversely affected by loss of mental acuity and irrationality. This would apply to most people in the United States". The original graphic can be seen in the Caltech Archives.

Demographic and Geographic Patterns

Blood lead levels are highest in countries where lead is added to petrol or gasoline, where lead is used in paint soldered products, in urban areas, in areas adjacent to high road traffic, and in developing countries. In Jamaica, 44% of children living near lead production facilities had BLLs above 25 µg/dL. In Albania, 98% of preschool children and 82% of schoolchildren had BLLs above 10 µg/dL; preschoolers living near a battery factory had average BLLs of 43 µg/dL. In China, 50% of children living in rural areas had BLLs above 10 µg/dL, and children living near sites of industry and high traffic had average BLLs ranging from 22 to 68 µg/dL.

BLL measurements from developed countries decreased markedly beginning in the late 1970s, when restrictions were placed upon lead use in gasoline, petrol, paint, soldering material and other products. In the United States, average BLLs measured among tens of thousands of subjects declined from 12.8 to 2.8 µg/dL between 1976 and 1991. In the 1990s, BLLs of children in Australia were measured to be 5 µg/dL, and 9 µg/dL in Barcelona, Spain.

In the United States, blood lead levels remain highest for children, for people in urban centers, for people of lower socioeconomic status, and for minorities.

Sources

Exposure to lead occurs through ingestion, inhalation, and dermal contact. Lead enters the bloodstream through exposure and elevates blood lead level that may result in lead poisoning or an elevated blood lead level. A major source of exposure to lead comes from inhalation. Factories and industries, vehicles exhausts, and even dust in the air that people breathe all have the potential of containing lead. Other major sources of lead exposure include ingestion and contact with products such as paint and soil that may contain lead. Many older claw-foot bathtubs have also been found to leach lead, especially when filled with warm bath water.

Health Effects

The Centers for Disease Control and Prevention (CDC) states "No safe blood lead level in children has been identified. Even low levels of lead in blood have been shown to affect IQ, ability to pay attention, and academic achievement. Effects of lead exposure cannot be corrected". "The absence of an identified BLL without deleterious effects, combined with the evidence that these effects appear to be irreversible, underscores the critical importance of primary prevention."

The most sensitive populations are infants, children, and pregnant women.

A child can drink a glass of water containing lead and absorb 50% of it. An adult might only retain

10% of the lead in that water. And once the lead is in the child's body, it reaches the brain through the not fully developed blood brain barrier. The body removes lead from blood and stores it in bone, but in children it subsequently leaves the bone more readily compared to adults. "Lead that has accumulated in a woman's bones is removed from her bones and passes freely from mother to child; maternal and fetal blood lead levels are virtually identical. Once in the fetal circulation, lead readily enters the developing brain through the immature blood–brain barrier".

"Lead is associated with a wide range of toxicity in children across a very broad band of exposures, down to the lowest blood lead concentrations yet studied, both in animals and people. These toxic effects extend from acute, clinically obvious, symptomatic poisoning at high levels of exposure down to subclinical (but still very damaging) effects at lower levels. Lead poisoning can affect virtually every organ system in the body. The principal organs affected are the central and peripheral nervous system and the cardiovascular, gastrointestinal, renal, endocrine, immune and haematological systems".

Adults who are exposed to a dangerous amount of lead can experience anemia, nervous system dysfunction, weakness, hypertension, kidney problems, decreased fertility, an increased level of miscarriages, premature deliveries, and low birth weight of their child.

Dietary Reference Intake

The Dietary Reference Intake (DRI) is a system of nutrition recommendations from the Institute of Medicine (IOM) of the National Academies (United States). It was introduced in 1997 in order to broaden the existing guidelines known as Recommended Dietary Allowances (RDAs). The DRI values differ from those used in nutrition labeling in the U.S. and Canada, which uses Reference Daily Intakes (RDIs) and Daily Values (%DV) based on outdated RDAs from 1968.

The DRI provides several different types of reference value:

- Estimated Average Requirements (EAR), expected to satisfy the needs of 50% of the people in that age group based on a review of the scientific literature.

- Recommended Dietary Allowances (RDA), the daily dietary intake level of a nutrient considered sufficient by the Food and Nutrition Board of the Institute of Medicine to meet the requirements of 97.5% of healthy individuals in each life-stage and sex group. The definition implies that the intake level would cause a harmful nutrient deficiency in just 2.5%. It is calculated based on the EAR and is usually approximately 20% higher than the EAR.

- Adequate Intake (AI), where no RDA has been established, but the amount established is somewhat less firmly believed to be adequate for everyone in the demographic group.

- Tolerable upper intake levels (UL), to caution against excessive intake of nutrients (like vitamin A) that can be harmful in large amounts. This is the highest level of daily nutrient consumption that is considered to be safe for, and cause no side effects in, 97.5% of healthy individuals in each life-stage and sex group. The definition implies that the intake level would cause a harmful nutrient excess in just 2.5%. The European Food Safety Authority (EFSA) has also established ULs which do not always agree with U.S. ULs. For example, zinc UL is 40 mg in U.S. and 25 mg in EFSA.

- Acceptable Macronutrient Distribution Ranges (AMDR), a range of intake specified as a percentage of total energy intake. Used for sources of energy, such as fats and carbohydrates.

The DRI is used by both the United States and Canada and is intended for the general public and health professionals. Applications include:

- Composition of diets for schools, prisons, hospitals or nursing homes

- Industries developing new food stuffs

- Healthcare policy makers and public health officials

History

The recommended dietary allowance (RDA) was developed during World War II by Lydia J. Roberts, Hazel Stiebeling, and Helen S. Mitchell, all part of a committee established by the United States National Academy of Sciences in order to investigate issues of nutrition that might "affect national defense".

The committee was renamed the Food and Nutrition Board in 1941, after which they began to deliberate on a set of recommendations of a standard daily allowance for each type of nutrient. The standards would be used for nutrition recommendations for the armed forces, for civilians, and for overseas population who might need food relief. Roberts, Stiebeling, and Mitchell surveyed all available data, created a tentative set of allowances for "energy and eight nutrients", and submitted them to experts for review (Nestle, 35).

The final set of guidelines, called RDAs for Recommended Dietary Allowances, were accepted in 1941. The allowances were meant to provide superior nutrition for civilians and military personnel, so they included a "margin of safety." Because of food rationing during the war, the food guides created by government agencies to direct citizens' nutritional intake also took food availability into account.

The Food and Nutrition Board subsequently revised the RDAs every five to ten years. In the early 1950s, United States Department of Agriculture nutritionists made a new set of guidelines that also included the number of servings of each food group in order to make it easier for people to receive their RDAs of each nutrient.

The DRI was introduced in 1997 in order to broaden the existing system of RDAs. DRIs were published over the period 1998 to 2001. In 2010, revised DRIs were published for calcium and vitamin D.

Current Recommendations

The current DRI values may differ from those used in nutrition labeling in the U.S. and Canada, which uses Reference Daily Intakes (RDIs) and Daily Values (%DV) based on RDAs from 1968.

Example of DRIs not matching DVs: the U.S. RDA for vitamin B12 is 2.4 µg/day whereas 100% DV is 6.0 µg/day.

Vitamins and Minerals

EARs, RDA/AIs and ULs for an average healthy 44-year-old male are shown below. Amounts and "ND" status for other age and gender groups, pregnant women, lactating women, and breastfeeding infants may be much different.

• "NE": EARs have not yet been established or not yet evaluated.

• "ND": ULs could not be determined, and it is recommended that intake from these nutrients be from food only, to prevent adverse effects.

Nutrient	EAR	RDA/AI	UL	Unit	Top Sources in Common Measures, USDA
Vitamin A	625	900	3000	µg	turkey and chicken giblets, liver, red capsicum, carrots, pumpkin, sweet potato
Vitamin C	75	90	2000	mg	guavas, oranges, grapefruits, frozen peaches,[i] bell peppers
Vitamin D	10	15	100	µg	fortified cereals, mushrooms, yeast, sockeye salmon, swordfish, rainbow trout, sardines, cod liver oil (also fortified foods and beverages)
Vitamin K	NE	120	ND	µg	kale, collards, spinach, broccoli, brussel sprouts, asparagus, prunes, green peas, blueberries, carrots
Vitamin B$_6$	1.1	1.3	100	mg	fortified cereals, chickpeas, sockeye salmon
α-tocopherol (Vitamin E)	12	15	1000	mg	fortified cereals, tomato paste, sunflower seeds
Biotin (B$_7$)	NE	30	ND	µg	whole grains, almonds, peanuts, beef liver, egg yolk, salmon
Calcium	800	1000	2500	mg	fortified cereals, collards, almonds, condensed cow's milk, cheese, figs, yogurt, milk
Chloride	NE	2300	3600	mg	table salt
Chromium	NE	35	ND	µg	broccoli, turkey ham, dried apricots, tuna, pineapple, grape juice
Choline	NE	550	3500	mg	egg yolk, meats, lecithin, beef liver, condensed milk, quinoa, salmon, cod
Copper	700	900	10000	µg	sesame seeds, sunflower seeds, oysters, lobster, cashews, dark chocolate, pearled barley, Brazil nuts, walnuts, peanuts, yellow peas, chickpeas
Cyanocobalamin (B$_{12}$)	2.0	2.4	ND	µg	fortified cereals, turkey, clams, beef, egg yolk, sardines, tuna fish, mackerel
Fluoride	NE	4	10	mg	public drinking water, where fluoridation is performed or natural fluorides are present
Folate (B$_9$)	320	400	1000	µg	leafy greens, enriched white rice, fortified cereals, enriched cornmeal
Iodine	95	150	1100	µg	iodized salt, kelp, cod
Iron	6	8	45	mg	cocoa powder, cashew nuts, white beans, turkey, dark chocolate
Magnesium	330	400	350[ii]	mg	buckwheat flour, rolled oats, spinach, almonds, dark chocolate, bulgur, quinoa

Manganese	NE	2.3	11	mg	oat bran, whole grain wheat flour, bulgur, rolled oats, brown rice, parboiled rice, dark chocolate
Molybdenum	34	45	2000	µg	legumes, grain products, nuts and seeds
Niacin (B$_3$)	12	16	35	mg	fortified cereals, yellowfin tuna, sockeye salmon, chicken meat
Pantothenic acid (B$_5$)	NE	5	ND	mg	fortified cereals, beef liver, shiitake mushrooms
Phosphorus	580	700	4000	mg	cornmeal, condensed milk, wheat flour, rolled oats, brown rice, bulgur, milk, meats
Potassium	NE	4700	ND	mg	potatoes, bananas, tomato paste, tomatoes, bell peppers, orange juice, beet greens, quinoa, rolled oats, bulgur, beans, peas, cashews, pistachio nuts
Riboflavin (B$_2$)	1.1	1.3	ND	mg	almonds, sesame seeds, spaghetti, beef liver, turkey
Selenium	45	55	400	µg	Brazil nuts, rockfish, tuna, beef, sardines, salmon, egg yolk, pearled barley, mackerel
Sodium	NE	1500	2300	mg	onion soup mix, miso, table salt, egg whites
Thiamin (B$_1$)	1.0	1.2	ND	mg	fortified cereals, enriched wheat flour, breadcrumbs
Zinc	9.4	11	40	mg	nuts, oysters, fortified cereals, beef, baked beans, oatmeal

EAR: Estimated Average Requirements; RDA: Recommended Dietary Allowances; AI: Adequate Intake; UL: Tolerable upper intake levels.

1. Vitamin C is added to frozen peaches to prevent darkening. Raw peaches and peaches preserved in syrup do not have a high vitamin C content.

2. The UL for magnesium represents extra intake from dietary supplements. High doses of magnesium from dietary supplements or medications often result in diarrhea that can be accompanied by nausea and abdominal cramping. There is no evidence of adverse effects from the consumption of naturally occurring magnesium in foods.

It is also recommended that the following substances not be added to food or dietary supplements. Research has been conducted into adverse effects, but was not conclusive in many cases:

Substance	RDA/AI	UL	units per day
Arsenic	–	ND	–
Silicon	–	ND	–
Vanadium	–	1.8	mg

Macronutrients

RDA/AI is shown below for males and females aged 40–50 years.

Substance	Amount (males)	Amount (females)	Top Sources in Common Measures
Water[i]	3.7 L/day	2.7 L/day	water, watermelon, iceberg lettuce
Carbohydrates	130 g/day	130 g/day	milk, grains, fruits, vegetables

Protein[ii]	56 g/day	46 g/day	meats, fish, legumes (pulses and lentils), nuts, milk, cheeses, eggs
Fiber	38 g/day	25 g/day	barley, bulgur, rolled oats, legumes, nuts, beans, apples,
Fat	20–35% of calories		oils, butter, lard, nuts, seeds, fatty meat cuts, egg yolk, cheeses
Linoleic acid, an omega-6 fatty acid (polyunsaturated)	17 g/day	12 g/day	sunflower seeds, sunflower oil, safflower oil,
alpha-Linolenic acid, an omega-3 fatty acid (polyunsaturated)	1.6 g/day	1.1 g/day	Linseed oil (Flax seed), salmon, sardines
Cholesterol	300 milligrams(mg)		chicken giblets, turkey giblets, beef liver, egg yolk
Trans fatty acids	As low as possible		
Saturated fatty acids	As low as possible while consuming a nutritionally adequate diet		coconut meat, coconut oil, lard, cheeses, butter, chocolate, egg yolk
Added sugar	No more than 25% of calories		foods that taste sweet but are not found in nature, such as sweets, cookies, cakes, jams, energy drinks, soda drinks, many processed foods

1.   Includes water from food, beverages, and drinking water.

2.   Based on 0.8 g/kg of body weight.

Calculating the RDA

The equations used to calculate the RDA are as follows:

"If the standard deviation (SD) of the EAR is available and the requirement for the nutrient is symmetrically distributed, the RDA is set at two SDs above the EAR:

If data about variability in requirements are insufficient to calculate an SD, a coefficient of variation (CV) for the EAR of 10 percent is assumed, unless available data indicate a greater variation in requirements. If 10 percent is assumed to be the CV, then twice that amount when added to the EAR is defined as equal to the RDA. The resulting equation for the RDA is then

This level of intake statistically represents 97.5 percent of the requirements of the population."

Standard of Evidence

In September 2007, the Institute of Medicine held a workshop entitled "The Development of DRIs 1994–2004: Lessons Learned and New Challenges." At that meeting, several speakers stated that the current Dietary Recommended Intakes (DRI's) were largely based upon the very lowest rank in the quality of evidence pyramid, that is, opinion, rather than the highest level – randomized controlled clinical trials. Speakers called for a higher standard of evidence to be utilized when making dietary recommendations.

Adherence

Nutrient	Percent of U.S. population ages 2+ meeting EAR in 2004
Protein	88.9%
Vitamin A	46.0%
Vitamin C	51.0%
Vitamin E	13.6%
Thiamin	81.6%
Riboflavin	89.1%
Niacin	87.2%
Vitamin B$_6$	73.9%
Folate	59.7%
Vitamin B$_{12}$	79.7%
Phosphorus	87.2%
Magnesium	43.0%
Iron	89.5%
Selenium	91.5%
Zinc	70.8%
Copper	84.2%
Calcium	30.9%
Fiber	8.0%
Potassium	7.6%
% calories from total fat <= 35%	59.4%
% calories from saturated fat < 10%	40.8%
Cholesterol intake < 300 mg	68.4%
Sodium intake <= 2,300 mg	29.9%

References

- Garcia, Javier; Colosio, Joëlle (2002). Air-quality indices : elaboration, uses and international comparisons. Presses des MINES. ISBN 2-911762-36-3.

- Rand G (1995). Fundamentals of Aquatic Toxicology: Effects, Environmental Fate, and Risk Assessment. Boca Raton, FL: CRC Press. ISBN 1-56032-091-5.

- Rama Lakshmi (17 October 2014). "India launches its own Air Quality Index. Can its numbers be trusted?". Washington Post. Retrieved 20 August 2015.

- "National Air Quality Index (AQI) launched by the Environment Minister AQI is a huge initiative under 'Swachh Bharat'". Retrieved 20 August 2015.

- "California". Community Nutrition Mapping Project. USDA Agricultural Research Service. "All U.S." column. Retrieved 6 Nov 2014.

- David Mintz (February 2009). Technical Assistance Document for the Reporting of Daily Air Quality – the Air Quality Index (AQI) (PDF). North Carolina: US EPA Office of Air Quality Planning and Standards. EPA-454/B-09-001. Retrieved 9 August 2012.

Study of Aquatic Toxicology

This chapter deals with the issue of toxicity rampant in aquatic organisms panning the freshwater, saltwater and sediment environments. The problem of bioaccumulation in aquatic organisms is a mammoth concern and this chapter provides extensive data on exposure systems, aquatic toxicity tests while also furnishing material on the toxicological effects of pollutants on aquatic organisms and how this indirectly affects human beings due to bioconcentration.

Aquatic Toxicology

Aquatic toxicology is the study of the effects of manufactured chemicals and other anthropogenic and natural materials and activities on aquatic organisms at various levels of organization, from subcellular through individual organisms to communities and ecosystems. Aquatic toxicology is a multidisciplinary field which integrates toxicology, aquatic ecology and aquatic chemistry.

This field of study includes freshwater, marine water and sediment environments. Common tests include standardized acute and chronic toxicity tests lasting 24–96 hours (acute test) to 7 days or more (chronic tests). These tests measure endpoints such as survival, growth, reproduction, that are measured at each concentration in a gradient, along with a control test. Typically using selected organisms with ecologically relevant sensitivity to toxicants and a well-established literature background. These organisms can be easily acquired or cultured in lab and are easy to handle.

History

While basic research in toxicology began in multiple countries in the 1800s, it was not until around the 1930s that the use of acute toxicity testing, especially on fish, was established. Over the next two decades, the effects of chemicals and wastes on non-human species became more of a public issue and the era of the *pickle-jar bioassays* began as efforts increased to standardize toxicity testing techniques. In the United States of America, the passage of the Federal Water Pollution Control Act of 1947 marked the first comprehensive legislation for the control of water pollution and was followed by the Federal Water Pollution Control Act in 1956. In 1962, public and governmental interests were renewed, in large part due to the publication of Rachel Carson's *Silent Spring*, and three years later the Water Quality Act was passed which directed states to develop water quality standards. Public awareness, as well as scientific and governmental concern, continued to grow throughout the 1970s and by the end of the decade research had expanded to include hazard evaluation and risk analysis. In the subsequent decades, aquatic toxicology has continued to expand and internationalize so that there is now a strong application of toxicity testing for environmental protection.

Aquatic Toxicity Tests

Aquatic toxicology tests (assays): toxicity tests are used to provide qualitative and quantitative data on adverse (deleterious) effects on aquatic organisms from a toxicant. Toxicity tests can be used to assess the potential for damage to an aquatic environment and provide a database that can be used to assess the risk associated within a situation for a specific toxicant. Aquatic toxicology tests can be performed in the field or in the laboratory. Field experiments generally refer to multiple species exposure and laboratory experiments generally refer to single species exposure. A dose response relationship is most commonly used with a sigmoidal curve to quantify the toxic effects at a selected end-point or criteria for effect (i.e. death or other adverse effect to the organism). Concentration is on the x-axis and percent inhibition or response is on the y-axis.

The criteria for effects, or endpoints tested for, can include lethal and sublethal effects.

There are different types of toxicity tests that can be performed on various test species. Different species differ in their susceptibility to chemicals, most likely due to differences in accessibility, metabolic rate, excretion rate, genetic factors, dieteary factors, age, sex, health and stress level of the organism. Common standard test species are the fathead minnow (Pimephales promelas), daphnids (*Daphnia magna, D. pulex, D. pulicaria, Ceriodaphnia dubia*), midge (Chironomus tentans, C. ruparius), rainbow trout (Oncorhynchus mykiss), sheepshead minnow (Cyprinodon variegatu), mysids (Mysidopsis), oyster (Crassotreas), scud (Hyalalla Azteca), grass shrimp (Palaemonetes pugio), mussels (Mytilus). As defined by ASTM, these species are routinely selected on the basis of availability, commercial, recreational, and ecological importance, past successful use, and regulatory use.

A variety of acceptable standardized test methods have been published. Some of the more widely accepted agencies to publish methods are: the American Public Health Association, U.S. Environmental Protection Agency, American Society for Testing and Materials, International Organization for Standardization, Environment Canada, and Organization for Economic Cooperation and Development. Standardized tests offer the ability to compare results between laboratories.

There are many kinds of toxicity tests widely accepted in the scientific literature and regulatory agencies. The type of test used depends on many factors: Specific regulatory agency conducting the test, resources available, physical and chemical characteristics of the environment, type of toxicant, test species available, laboratory vs. field testing, end-point selection, and time and resources available to conduct the assays are some of the most common influencing factors on test design.

Exposure Systems

Exposure systems are four general techniques the controls and test organisms are exposed to the dealing with treated and diluted water or the test solutions.

Static- a static test exposes the organism in still water. The toxicant is added to the water in order to obtain the correct concentrations to be tested. The control and test organisms are placed in the test solutions and the water is not changed for the entirety of the test.

Recirculation- a recirculation test exposes the organism to the toxicant in a similar manner as the static test, except that the test solutions are pumped through an apparatus (i.e. filter) to maintain water quality, but not reduce the concentration of the toxicant in the water. The water is circulated through the test chamber continuously, similar to an aerated fish tank. This type of test is expensive and it is unclear whether or not the filter or aerator has an effect on the toxicant.

Renewal- a renewal test also exposes the organism to the toxicant in a similar manner as the static test because it is in still water. However, in a renewal test the test solution is renewed periodically (constant intervals) by transferring the organism to a fresh test chamber with the same concentration of toxicant.

Flow-through- a flow through test exposes the organism to the toxicant with a flow into the test chambers and then out of the test chambers. The once-through flow can either be intermittent or continuous. A stock solution of the correct concentrations of contaminant must be previously prepared. Metering pumps or diluters will control the flow and the volume of the test solution, and the proper proportions of water and contaminant will be mixed.

Types of Tests

Acute tests are short-term exposure tests (hours or days) and generally use lethality as an endpoint. In acute exposures, organisms come into contact with higher doses of the toxicant in a single event or in multiple events over a short period of time and usually produce immediate effects, depending on absorption time of the toxicant. These tests are generally conducted on organisms during a specific time period of the organism's life cycle, and are considered partial life cycle tests. Acute tests are not valid if mortality in the control sample is greater than 10%. Results are reported in EC50, or concentration that will affect fifty percent of the sample size.

Chronic tests are long-term tests (weeks, months years), relative to the test organism's life span (>10% of life span), and generally use sub-lethal endpoints. In chronic exposures, organisms come into contact with low, continuous doses of a toxicant. Chronic exposures may induce effects to acute exposure, but can also result in effects that develop slowly. Chronic tests are generally considered full life cycle tests and cover an entire generation time or reproductive life cycle ("egg to egg"). Chronic tests are not considered valid if mortality in the control sample is greater than 20%. These results are generally reported in NOECs (No observed effects level) and LOECs (Lowest observed effects level).

Early life stage tests are considered as subchronic exposures that are less than a complete reproductive life cycle and include exposure during early, sensitive life stages of an organism. These exposures are also called critical life stage, embryo-larval, or egg-fry tests. Early life stage tests are not considered valid if mortality in the control sample is greater than 30%.

Short-term sublethal tests are used to evaluate the toxicity of effluents to aquatic organisms. These methods are developed by the EPA, and only focus on the most sensitive life stages. Endpoints for these test include changes in growth, reproduction and survival. NOECs, LOECs and EC50s are reported in these tests.

Bioaccumulation tests are toxicity tests that can be used for hydrophobic chemicals that may accumulated in the fatty tissue of aquatic organisms. Toxicants with low solubilities in water generally can be stored in the fatty tissue due to the high lipid content in this tissue. The storage of these toxicants within the organism may lead to cumulative toxicity. Bioaccumulation tests use bioconcentration factors (BCF) to predict concentrations of hydrophobic contaminants in organisms. The BCF is the ratio of the average concentration of test chemical accumulated in the tissue of the test organism (under steady state conditions) to the average measured concentration in the water.

Freshwater tests and saltwater tests have different standard methods, especially as set by the regulatory agencies. However, these tests generally include a control (negative and/or positive), a geometric dilution series or other appropriate logarithmic dilution series, test chambers and equal numbers of replicates, and a test organism. Exact exposure time and test duration will depend on type of test (acute vs. chronic) and organism type. Temperature, water quality parameters and light will depend on regulator requirements and organism type.

Effluent toxicity tests are tests conducted under the Clean Water Act, National Pollutant Discharge Elimination System (NPDES) permit program and are used by dischargers of contaminated effluent to monitor the quality of effluent into receiving waters. Acute Effluent Toxicity Tests are used to monitor the quality of industrial effluent monthly using acute toxicity tests. Effluent is used to perform static-acute multi concentration toxicity tests with *Ceriodaphnia dubia* and *Pimephales promelas*. The test organisms are exposed for 48 hours under static conditions with five concentrations of the effluent. Short-term Chronic Effluent Toxicity Tests are used to monitor the quality of municipal wastewater treatment plants effluent quarterly using short-term chronic toxicity tests. The goal of this test is to ensure that the wastewater is not chronically toxic. The major deviation in the short-term chronic effluent toxicity tests and the acute effluent toxicity tests is that the short-term chronic test lasts for seven days and the acute test lasts for 48 hours.

Sediment Tests

At some point most chemicals originating from both anthropogenic and natural sources accumulate in sediment. For this reason, sediment toxicity can play a major role in the adverse biological effects seen in aquatic organisms, especially those inhabiting benthic habitats. A recommended approach for sediment testing is to apply the Sediment Quality Triad (SQT) which involves simultaneously examining sediment chemistry, toxicity, and field alterations so that more complete information can be gathered. Collection, handling, and storage of sediment can have an effect on bioavailability and for this reason standard methods have been developed to suit this purpose.

Toxicological Effects

Toxicity can be broken down into two broad categories of direct and indirect toxicity. Direct toxicity results from a toxicant acting at the site of action in or on the organism. Indirect toxicity occurs with a change in the physical, chemical, or biological environment.

Lethality is most common effect used in toxicology and used as an endpoint for acute toxicity tests. While conducting chronic toxicity tests sublethal effects are endpoints that are looked at. These endpoints include behavioral, physiological, biochemical, histological changes.

There are a number of effects that occur when an organism is simultaneously exposed to two or more toxicants. These effects include additive effects, synergistic effects, potentiation effects, and antagonistic effects. An additive effect occurs when combined effect is equal to a combination or sum of the individual effects. A synergistic effect occurs when the combination of effects is much greater than the two individual effects added together. Potentiation is an effect that occurs when an individual chemical has no effect is added to a toxicant and the combination has a greater effect than just the toxicant alone. Finally, an antagonistic effect occurs when a combination of chemicals has less of an effect than the sum of their individual effects.

Important Aquatic Toxicology Resources

- American Society for Testing and Materials (ASTM International) – A consensus organization, representing 135 countries, that develops and delivers international voluntary standard methods for aquatic toxicity testing.

- Standard Methods for the Examination of Water and Wastewater – A compilation of techniques for the examination of water, jointly published by the American Public Health Association (APHA), the American Water Works Association (AWWA), and the Water Pollution Control Federation (WPCF).

- Ecotox – A database maintained by the U.S. Environmental Protection Agency (EPA) that offers single chemical toxicity information for both aquatic and terrestrial purposes.

- Society of Environmental Toxicology and Chemistry (SETAC) – A nonprofit, worldwide society working to promote scientific research to further our understanding of environmental stressors, environmental education, and the use of science in environmental policy.

- United States Environmental Protection Agency (EPA) – A federal agency working to protect human and environmental health. Among many other functions, the U.S. EPA produces guidance manuals outlining aquatic toxicity test procedures.

- Organisation for Economic Co-operation and Development (OECD) – A forum for governments to work together to promote policies for the betterment of people's social and economic well-being around the world. One way in which they accomplish this is through the development of aquatic toxicity test guidelines.

- Environment Canada (EC) – A diverse organization working to protect Canada's water resources and the natural environment through the coordination of environmental policies and programs with the federal government.

Terminology

- Median Lethal Concentration (LC50) – The chemical concentration that is expected to kill 50% of a group of organisms.

- Median Effective Concentration (EC50) – The chemical concentration that is expected to have one or more specified effects in 50% of a group of organisms.

- Critical Body Residue (CBR) – An approach that routinely examines whole-body chemical concentrations of an exposed organism that is associated with an adverse biological response.

- Baseline toxicity – Refers to narcosis which is a depression in biological activity due to toxicants being present in the organism.

- Biomagnification – The process by which the concentration of a chemical in the tissues of an organism increases as it passes through several levels in the food web.

- Lowest Observed Effect Concentration (LOEC) – The lowest test concentration that has a statistically significant effect over a specified exposure time.

- No Observed Effect Concentration (NOEC) – The highest test concentration for which no effect is observed relative to a control over a specified exposure time.

- Maximum Acceptable Toxicant Concentration (MATC) – An estimated value that represents the highest "no-effect" concentration of a specific substance within the range including the NOEC and LOEC.

- Application Factor (AF) – An empirically derived "safe" concentration of a chemical.

- Biomonitoring – The consistent use of living organisms to analyze environmental changes over time.

- Effluent – Liquid, industrial discharge that usually contain varying chemical toxicants.

- Quantitative Structure-Activity Relationship (QSAR) – A method of modeling the relationship between biological activity and the structure of organic chemicals.

- Mode of Action – A set of common behavioral or physiological signs that represent a type of adverse response.

- Mechanism of Action – The detailed events that take place at the molecular level during an adverse biological response.

- KOW – The octanol-water partition coefficient which represents the ratio of the concentration of octanol to the concentration of chemical in the water.

- Bioconcentration Factor (BCF) – The ratio of the average chemical concentration in the tissues of the organism under steady-state conditions to the average chemical concentration measured in the water to which the organisms are exposed.

All terms were derived from Rand.

Significance to Regulatory World

In the United States aquatic toxicology plays an important role in the NPDES wastewater permit program. In addition to analytical testing for known pollutants, aquatic, whole effluent toxicity tests have been standardized and are performed routinely as a tool for evaluating the potential harmful effects of effluents discharged into surface waters.

For the Clean Water Act under United States Environmental Protection Agency there are water quality criteria and water quality standards derived from aquatic toxicity tests.

Sediment Quality Guidelines

While sediment quality guidelines are not meant for regulation, they provide a way to rank and compare sediment quality developed by National Oceanic and Atmospheric Administration(-NOAA). These sediment quality guidelines are summarized in NOAA's Screening Quick Reference Tables (SQuiRT) for many different chemicals.

Bioconcentration

Bioconcentration is the accumulation of a chemical in or on an organism when the source of chemical is solely water. Bioconcentration is a term that was created for use in the field of aquatic toxicology. Bioconcentration can also be defined as the process by which a chemical concentration in an aquatic organism exceeds that in water as a result of exposure to a waterborne chemical.

There are several ways in which to measure and assess bioaccumulation and bioconcentration. These include: octanol-water partition coefficients (K_{OW}), bioconcentration factors (BCF), bioaccumulation factors (BAF) and biota-sediment accumulation factor (BSAF). Each of these can be calculated using either empirical data or measurements as well as from mathematical models. One of these mathematical models is a fugacity-based BCF model developed by Don Mackay.

Bioconcentration factor can also be expressed as the ratio of the concentration of a chemical in an organism to the concentration of the chemical in the surrounding environment. The BCF is a measure of the extent of chemical sharing between an organism and the surrounding environment.

In surface water, the BCF is the ratio of a chemical's concentration in an organism to the chemical's aqueous concentration. BCF is often expressed in units of liter per kilogram (ratio of mg of chemical per kg of organism to mg of chemical per liter of water). BCF can simply be an observed ratio, or it can be the prediction of a partitioning model. A partitioning model is based on assumptions that chemicals partition between water and aquatic organisms as well as the idea that chemical equilibrium exists between the organisms and the aquatic environment in which it is found

Calculation

Bioconcentration can be described by a bioconcentration factor (BCF), which is the ratio of the chemical concentration in an organism or biota to the concentration in water:

$$BCF = \frac{Concentration_{Biota}}{Concentration_{Water}}$$

Bioconcentration factors can also be related to the octanol-water partition coefficient, K_{OW}. The octanol-water partition coefficient (K_{OW}) is correlated with the potential for a chemical to bioaccumulate in organisms; the BCF can be predicted from log K_{OW}, via computer programs based on structure activity relationship (SAR) or through the linear equation:

$$logBCF = mlogK_{OW} + b$$

Where:

$$K_{OW} = \frac{Concentration_{octanol}}{Concentration_{water}} = \frac{C_O}{C_W} \text{ at equilibrium}$$

Fugacity Capacity

Fugacity and BCF relate to each other in the following equation:

$$Z_{Fish} = \frac{P_{Fish} \times BCF}{H}$$

where Z_{Fish} is equal to the Fugacity capacity of a chemical in the fish, P_{Fish} is equal to the density of the fish (mass/length³), BCF is the partition coefficient between the fish and the water (length³/mass) and H is equal to the Henry's law constant (Length²/Time²)

Regression Equations for Estimations in Fish

Equation	Chemicals Used to obtain equation	Species Used
$logBCF = 0.76logKow - 0.23$	84	Fathead Minnow, Bluegill Sunfish, Rainbow Trout, Mosquitofish
$logBCF = logKow - 1.32$	44	Various
$logBCF = 2.791 - 0.564logS (S = watersolubility)$	36	Brook trout, Rainbow trout, Bluegill Sunfish, Fathead minnow, Carp
$logBCF = 3.41 - 0.508logS$	7	Various
$logBCF = 1.119logKoc - 1.579$	13	Various

Uses

Regulatory Uses

Through the use of the PBT Profiler and using criteria set forth by the United States Environmental Protection Agency under the Toxic Substances Control Act (TSCA), a substance is considered to be not bioaccumulative if it has a BCF less than 1000, bioaccumulative if it has a BCF from 1000–5000 and very bioaccumulative if it has a BCF greater than 5,000.

The thresholds under REACH are a BCF of > 2000 l/kg bzw. for the B and 5000 l/kg for vB criteria.

Applications

A bioconcentration factor greater than 1 is indicative of a hydrophobic or lipophilic chemical. It is an indicator of how probable a chemical is to bioaccumulate. These chemicals have high lipid affinities and will concentrate in tissues with high lipid content instead of in an aqueous environment like the cytosol. Models are used to predict chemical partitioning in the environment which in turn allows the prediction of the biological fate of lipophilic chemicals.

Equilibrium Partitioning Models

Based on an assumed steady state scenario, the fate of a chemical in a system is modeled giving predicted endpoint phases and concentrations.

It needs to be considered that reaching steady state may need a substantial amount of time as estimated using the following equation (in hours).

$$t_{eSS} = 0.00654 \cdot K_{OW} + 55.31$$

For a substance with a K_{OW} of 4, it thus takes approximately five days to reach effective steady state. For a K_{OW} of 6, the equilibrium time increases to nine months.

Fugacity Models

Fugacity is another predictive criterion for equilibrium among phases that has units of pressure. It is equivalent to partial pressure for most environmental purposes. It is the absconding propensity of a material. BCF can be determined from output parameters of a fugacity model and thus used to predict the fraction of chemical immediately interacting with and possibly having an effect on an organism.

Food Web Models

If organism-specific fugacity values are available, it is possible to create a food web model which takes trophic webs into consideration. This is especially pertinent for conservative chemicals that are not easily metabolized into degradation products. Biomagnification of conservative chemicals such as toxic metals can be harmful to apex predators like orca whales, osprey, and bald eagles.

Applications to Toxicology

Predictions

Bioconcentration factors facilitate predicting contamination levels in an organism based on chemical concentration in surrounding water. BCF in this setting only applies to aquatic organisms. Air breathing organisms do not take up chemicals in the same manner as other aquatic organisms. Fish, for example uptake chemicals via ingestion and osmotic gradients in gill lamellae.

When working with benthic macroinvertebrates, both water and benthic sediments may contain chemical that affects the organism. Biota-sediment accumulation factor (BSAF) and biomagnification factor (BMF) also influence toxicity in aquatic environments.

BCF does not explicitly take metabolism into consideration so it needs to be added to models at other points through uptake, elimination or degradation equations for a selected organism.

Body Burden

Chemicals with high BCF values are more lipophilic, and at equilibrium organisms will have greater concentrations of chemical than other phases in the system. Body burden is the total amount of chemical in the body of an organism, and body burdens will be greater when dealing with a lipophilic chemical.

Biological Factors

In determining the degree at which bioconcentration occurs biological factors have to be kept in mind.The rate at which an organism is exposed through respiratory surfaces and contact with dermal surfaces of the organism, competes against the rate of excretion from an organism. The rate of excretion is a loss of chemical from the respiratory surface, growth dilution, fecal excretion, and metabolic biotransformation. Growth dilution is not an actual process of excretion but due to the mass of the organism increasing while the contaminant concentration remains constant dilution occurs.

The interaction between inputs and outputs is shown here:

$$\frac{dC_B}{dt} = (k_1 C_{WD}) - (k_2 + k_E + k_M + k_G)C_B$$

The variables are defined as:
C_B is the concentration in the organism ($g*kg^{-1}$). t represents a unit of time (d^{-1}). k_1 is the rate constant for chemical uptake from water at the respiratory surface ($L*kg^{-1}*d^{-1}$). C_{WD} is the chemical concentration dissolved in water ($g*L^{-1}$). k_2, k_E, k_G, k_B are rate constants that represent excretion from the organism from the respiratory surface, fecal excretion, metabolic transformation, and growth dilution (d^{-1}).

Static variables influence BCF as well. Because organisms are modeled as bags of fat, lipid to water ratio is a factor that needs to be considered. Size also plays a role as the surface to volume ratio influence the rate of uptake from the surrounding water. The species of concern is a primary factor in influencing BCF values due to it determining all of the biological factors that alter a BCF.

Environmental Parameters

Temperature

Temperature may affect metabolic transformation, and bioenergetics. An example of this is the movement of the organism may change as well as rates of excretion. If a contaminant is ionic, the change in pH that is influenced by a change in temperature may also influence the bioavailability

Water Quality

The natural particle content as well as organic carbon content in water can affect the bioavailability. The contaminant can bind to the particles in the water, making uptake more difficult, as well as become ingested by the organism. This ingestion could consist of contaminated particles which would cause the source of contamination to be from more than just water.

References

- Rand, Gary M.; Petrocelli, Sam R. (1985). Fundamentals of aquatic toxicology: Methods and applications. Washington: Hemisphere Publishing. ISBN 0-89116-382-4.

- Landis WG, Sofield RM, Yu MH (2011). Introduction to Environmental Toxicology: Molecular Structures to Ecological Landscapes (Fourth ed.). Boca Raton, FL: CRC Press. pp. 117–162. ISBN 978-1-4398-0410-0.

- Hemond, Harold (2000). Chemical Fate and Transport in the Environment. San Diego, CA: Elsevier. pp. 156–157. ISBN 978-0-12-340275-2.

- Arnot, Jon; Frank A.P.C. Gobas (13 December 2006). "A review of bioconcentration factor (BCF) and bioaccumulation factor (BAF) assessments for organic chemicals in aquatic organisms" (PDF). NRC Canada. 12: 257–297. doi:10.1139/A06-005. Retrieved 7 June 2012.

- "About the Organisation for Economic Co-operation and Development (OECD)" Organisation for Economic Co-operation and Development, Retrieved 2012-06-07

- EPA. "Category for Persistent, Bioaccumulative, and Toxic New Chemical Substances". Federal Register Environmental Documents. USEPA. Retrieved 3 June 2012.

Permissions

Index

A

Accidental, 55, 95, 129, 135, 137

Acute Toxicity, 75, 140, 143, 156, 183, 204, 207

Air Quality Index, 166-168, 171, 174-177, 179, 203

Anticholinesterase Compounds, 137

Aquatic Toxicology, 123, 156, 179, 181, 203-205, 208-210

Arsenic Contamination Of Groundwater, 5, 13

B

Bacterial, 12, 86, 164, 189

Bee Kill Rate Per Hive, 145

Bioaccumulation, 1, 3, 5-6, 19-20, 26, 58, 99, 124-125, 129, 136, 140, 153, 158-160, 162, 184, 204, 206-207, 210, 214

Bioanalysis, 189, 191

Bioanalytical Organisations, 191

Bioconcentration, 5, 183, 204, 207, 209-212, 214

Biological Half-life, 5, 90

Biomagnification, 1, 3, 19-20, 25, 125, 158-160, 209, 212

Blood Lead Level, 31-32, 34-36, 40, 43, 45-48, 195-197

C

Cadmium, 2, 6-9, 12, 54-56

Cadmium Poisoning, 54, 56

Carbon Monoxide Poisoning, 5, 101, 103, 105-108, 110-111

Carcinogenicity, 75, 120, 140

Cardiovascular System, 35

Central Nervous System, 19, 32, 34-36, 44, 47-49, 101, 103, 106, 129, 182, 196

Cholinesterase, 137

Chromium, 6-9, 11, 122, 200

Chronic Effects, 87, 139, 143

Chronic Poisoning, 5, 34, 56, 103

Chronic Toxicity, 75, 156, 161, 179, 204, 207

Colony Collapse Disorder, 135, 142-144, 151

Combustion Products, 94, 96-97

Contamination Sources, 7

Contamination Specific Nations, 14

Context-dependent Pathogenicity, 163

Copper, 2, 6-7, 9, 13, 38, 40, 51, 158, 200, 203

Culture, 111, 135, 139, 171

Cumulative, 44, 49, 143, 207

Cytochrome Oxidase, 101, 104-106, 110

D

Ddt, 1, 3, 5, 68-81, 127-128, 130, 134, 136, 154, 157, 161, 163

Deepwater Horizon Oil Spill, 57

Diagnosis, 31-32, 43-44, 71, 76-77, 106-108, 137

Diazinon, 147, 157

Dietary Intake, 19, 74, 198

Dietary Reference Intake, 166, 198

Dietary Sources, 99

Direct Approach, 90, 186-187

Disparate Impacts, 26

E

Early Life Stage Test, 166, 179

Ecological Death, 153, 156-158

Endocrine Disruption, 129-130

Environmental Hazard, 153-154

Enzymes, 8, 42, 69, 112, 119, 157, 160

Exposure Assessment, 184, 186, 188-189

Extraction, 13, 114, 116, 190

Exxon Valdez Oil Spill, 57-59

F

Fallout Protection, 67

Fungal, 165

G

Gastrointestinal Tract, 34, 44, 46, 99, 125

Global Fallout, 60

H

Health Effects, 3, 26, 31-32, 35, 41, 49, 79, 84, 91, 96, 99-100, 118, 123, 127, 129-130, 132, 166-167, 173, 177, 184, 197

Health Effects of Pesticides, 132

Hemoglobin, 34, 42, 101, 104-108, 110, 112

Histological Effects, 157

I

Indirect Approach, 186

Isomers, 69, 114, 121, 128

L

Large-scale Water Treatment, 17

Lead, 2, 5-12, 26, 28, 31-55, 57-59, 67, 86, 93-95, 101-102, 106, 119, 121, 123-124, 130, 136, 146, 154, 156-159, 161, 179, 187, 195-198, 207

Lead Poisoning, 5, 7-8, 11-12, 31-35, 37-38, 40, 43-54, 123, 197-198

Lead-containing Products, 37, 39

Levels of Contamination, 22

Local Fallout, 60-64

M

Mechanism of Insecticide Action, 69

Mercury, 2, 5-10, 19-20, 22, 24-31, 36, 97-100, 154, 158, 161-162

Mercury in Fish, 19-20, 25-26

Meteorological, 63-64

Methodology, 97, 179-180

Methylcyclopentadienyl

Manganese Tricarbonyl, 93

Methylmercury, 5, 10, 19, 24, 28-30, 97-100

Modes of Toxic Action, 166, 181-183

Myoglobin, 101, 104-105

N

Neonicotinoids, 142-143, 151

Neurons, 42-43, 69

Nomenclature, 113

Nuclear Fallout, 5, 60-61, 66-67

O

Occupational, 31, 37, 45, 52, 56, 76, 89-90, 95-96, 118, 120, 129, 132, 135-137, 184, 188-189

Occupational Exposure, 31, 37, 95, 118, 120, 132, 135, 188-189

Organochlorines, 75, 136

Origins of Mercury Pollution, 24

P

Pathophysiology, 41, 53, 104, 136

Pentachlorophenol, 157

Persistent Organic Pollutant, 73, 124, 153

Pesticide Poisoning, 134-138

Pesticide Toxicity, 142-143

Pesticide Toxicity to Bees, 142

Pesticides, 2-3, 5, 26, 39, 48, 68, 71-72, 78, 124, 127, 130, 132-145, 151-152, 154, 159, 182

Pesticides Formulations, 145

Physiological Effects, 156

Polycyclic Aromatic Hydrocarbon, 57, 112

Prevention, 31-32, 45-46, 75, 78, 80, 101, 108, 134, 137, 165, 195, 197

Prionic, 165

Production, 3, 18-20, 25, 38, 40, 42, 49, 56, 58, 68-69, 79-80, 98, 104, 116-117, 119, 124, 127-128, 132, 158, 162, 182, 197

R

Receptor-based Approach, 184-185

Regulatory Uses, 181, 211

Remediation, 12, 46

Reproductive System, 3, 35, 75, 129-130

Residential, 52, 58, 78, 107, 109, 132, 136, 147

Routes of Exposure, 31, 43, 59, 95, 185

S

Seafood Consumption, 29

Seafood Consumption Benefits, 29

Small-scale Water Treatment, 17

Society, 52, 74, 111, 135, 139, 191, 205, 208

Speciation of Arsenic

Compounds in Water, 14

Specific Toxicants, 1, 160, 180, 182

Sub-lethal Toxicity, 143

Suicidal, 135

T

Toxic Equivalency Factor, 191-192

Toxic Heavy Metal, 6-8, 13

Toxicity Class, 139-141

Toxicity Class By Jurisdiction, 140

V

Viral, 44, 86, 108, 164

Volatility, 67, 115, 160, 162

W

Water Purification, 14, 17

World Health Organization, 11, 32, 71, 73, 100, 109, 134-135, 140, 194-196

www.ingramcontent.com/pod-product-compliance
Lightning Source LLC
Chambersburg PA
CBHW082040190326
41458CB00010B/3421